International Conference on Alternative Fuels and Electric Vehicles 2021 (ICAFEV 2021)

IOP Conference Series: Earth and Environmental Science
Volume 1042

Online
9-10 December 2021

ISBN: 978-1-7138-5820-1
ISSN: 1755-1307

Printed from e-media with permission by:

Curran Associates, Inc.
57 Morehouse Lane
Red Hook, NY 12571

Some format issues inherent in the e-media version may also appear in this print version.

Institute of Physics
Dirac House, Temple Back
Bristol BS1 6BE UK

Phone: 44 1 17 929 7481
Fax: 44 1 17 920 0979

techtracking@iop.org

Additional copies of this publication are available from:

Curran Associates, Inc.
57 Morehouse Lane
Red Hook, NY 12571 USA
Phone: 845-758-0400
Fax: 845-758-2633
Email: curran@proceedings.com
Web: www.proceedings.com

TABLE OF CONTENTS

Preface: International Conference on Alternative Fuels and Electric Vehicles 2021 (ICAFEV 2021)
Kumar D. Sapate, Sandip A. Kale

Peer Review Statement

Influence of Different Biodiesel Fuels on the PCCI-DI Combustion Concept for Constant Speed Genset Application 1
Hemant A Kinikar, A B Kanasepatil, S S Thipse

Investigation of Mahua, Pongamia and Waste Cooking Oil Blends as Biodiesel Properties 14
Malati Chanagond, Prashant Puthani, Ramesh Jeeragal, Suresh Doddi

Response Surface Methodology to Predict the Effects of Graphene Nanoplatelets Addition to Diesel Engine Performance 23
Sarbani Daud, Mohd Adnin Hamidi, Rizalman Mamat

Influence of Simarouba Glauca (Laxmi Taru) Biodiesel on Engine Performance and Emissions of Conventional Compression Ignition Engine 31
Prashant Puthani, Ramesh Jeeragal, Amaresha T Patil, Malati Chanagond

Characterization of Chia Seed Oil Methyl Ester as an Alternate Fuel for Diesel Engine 41
S Manjunath, Ramakrishna N Hedge

Design and Fabrication of a Partial Flow Dilution Tunnel for Particulate Mass Sampling in an Ethanol Blended CI Engine 48
Atul Kumar, Tomesh Kumar Sahu, Pravesh Chandra Shukla

Studies on the Catalytic Hydrothermal Liquefaction of Sugarcane Bagasse to Bio-Oil, Over Cobalt Oxide 55
Gopalakrishnan Govindasamy, Rohit Sharma, Sunu Subramanian

Bio-Oil by the Hydrothermal Liquefaction of Municipal Sewage Sludge Using Hierarchical Fe-Co-ZSM-5 Catalysts 61
Gopalakrishnan Govindasamy, Praveen Kumar J, Ramatejaswi Pasupula

Environmentally Balanced Preconditions for Use of Logging Residues in Forests of Ukrainian Carpathians for Energy Production 69
R Vasylyshyn, Lakyda, O Soshenskyi, M Lakyda, O Melnyk, V Slusarchuk

Energy Potential of Woody Biomass in Ukraine's Forests and Prospects for Its Utilization as an Alternative Energy Source 75
R Vasylyshyn, I Lakyda, Yu Yurchuk, M Lakyda, O Melnyk, R Bondarchuk

The Performance Analysis of Modified Generic Model for Lithium-Ion Batteries 81
Shubha Rao K, Karanam Vasudha

An Overview of Lithium-Ion Batteries for Electric Mobility and Energy Storage Applications 89
Ganesh Sankaran, S. Venkatesan

Exploring the State of Solar Photovoltaic Decentralization in Ghana: Trends and Success Factors 100
David M Sackey, De-Graft Owusu-Manu, Amevi Acakpovi, Frank K Appiah, David J Edwards, Reza M Hosseini

Optimal Sizing and Techno-Economic Analysis of a Hybrid Solar PV/Wind/Diesel Generator System ... 112
Godfred Tay, Amevi Acakpovi, Patrick Adjei, George K. Aggrey, Robert Sowah, Daniel Kofi, Maxwell Afonope, Mustapha Sulley

Numerical Analysis Using SolTrace Software to Determine the Size and Position of a Central Receiver for Ganged Heliostat System ... 127
Lalit Jadhav, Prem Gadiya, Jwalandhar Girnar, Harshvardhan Bhusal, Prakash Gadhe

Techno-Economic Assessment of Grid Connected Solar PV/Wind Hybrid System ... 142
Mustapha Sulley, Amevi Acakpovi, Patrick Adjei, David M. Sackey, Felix Offei, Maxwell Afonope, Daniel Kofi, Godfred Tay

Synthesis of ZnO Nanoparticles Through a Simple Wet Chemical Precipitation Method ... 157
Gem Taka, T D Das

Innovative Design of Grid Connected Solar/Diesel Hybrid System Using RETScreen Software ... 163
Daniel Kofi, Amevi Acakpovi, Patrick Adjei, Robert Sowah, George K. Aggrey, Godfred Tay, Mustapha Sulley, Maxwell Afonope

Economic-Environmental Impacts on Reduction of Energy Consumption in Technical High Schools During the COVID-19 Pandemic: Case of Manzanillo, Mexico ... 173
V D Cuevas-Rodríguez, G M Rodríguez-Torres, M Morales-Máximo, L B López-Sosa

Author Index

IOP Conf. Series: Earth and Environmental Science 1042 (2022) 011001 doi:10.1088/1755-1315/1042/1/011001

Preface: International Conference on Alternative Fuels and Electric Vehicles 2021 (ICAFEV 2021)

Kumar D. Sapate[1], Sandip A. Kale[2]

[1] Alard College of Engineering and Management, Pune, India

[2] Technology Research and Innovation Centre, Pune, India

International Conference on Alternative Fuels and Electric Vehicles 2021 (ICAFEV 2021) was hosted by the Alard College of Engineering and Management, Pune, India in association with Technology Research and Innovation Centre, Pune, India and Accra Technical University, Accra, Ghana on 09-10 December 2021 and conducted in virtual mode. This international meeting provided a platform to various worldwide researchers, and academicians engaged in the field of alternative fuels, electric vehicles and renewable energy with the motive of exploring the research to drive the World towards a sustainable future. ICAFEV 2021 received papers from authors of eleven countries such as Cameroon, Chile, Ghana, India, Malaysia, Mexico, Thailand, Turkey, Ukraine, Australia and United Kingdom.

This volume is a compilation of research papers presented in three tracks, 1) Alternative Fuels, 2) Electric Vehicles and 3) Renewable Energy. We feel, these papers will provide a valuable contribution in the field and readers will receive this content for their future research activities.

The conference organizers and associate organizers are thankful to all authors, technical committee members, advisory committee members and organizing committee members for contributing actively and whole heartily in successful completion of the conference throughout the various stages. We also express out gratitude towards the IOP publishing for providing support to publish the ICAFEV 2021 papers in IOP Conference Series: Earth and Environmental Science.

ICAFEV 2021 Committees

Patrons
- Dr. L. R. Yadav, Chairman, Alard Group of Institutes, Pune, India
- Prof. Felix Kutsanedzie, Dean, Faculty of Engineering, Accra Technical University

Conference Chairs
- Prof. Dr. Kumar D. Sapate, Alard College of Engineering and Management, Pune, India
- Prof. Dr. Amevi Acakpovi, Electrical Engineering Department, Accra Technical University

Publication Chair
- Dr. Sandip A. Kale, Technology Research and Innovation Centre, Pune, India

Technical Committee
- Dr. A. S. N. Chakravarthy, JNTUK, University College of Engineering, Vizianagaram, India
- Dr. A.P. Shah, Rajarambapu Institute of Technology, Maharashtra, India
- Dr. Avik Bhattacharya, Indian Institute of Technology Roorkee, India
- Dr. B. S. Gandhare, SKN Sinhgad College of Engineering, Pandharpur, Maharashtra, India
- Dr. Dnyandeo K. Shedge, AISSMS Institute of Information Technology, Pune, India
- Dr. E. Anil Kumar, Indian Institute of Technology Tirupati, India
- Dr. Geetha Sivakumar, Lovely Professional University, Phagwara, India
- Dr. Glenn Kwabena Gyimah, Accra Technical University, Accra, Ghana
- Dr. J. M. Mallikarjuna, Indian Institute of Technology Madras, Chennai, India
- Dr. L. M. Jugulkar, Rajarambapu Institute of Technology, Maharashtra, India
- Dr. N. A. Rawabawale, College of Engineering, Ambajogai, Maharashtra, India
- Dr. Nana Yaw Asabere, Accra Technical University, Accra, Ghana
- Dr. Peter Nyanor, Accra Technical University, Accra, Ghana
- Dr. Prakash Gadhe, Maharashtra Institute of Technology, Pune, Maharashtra, India
- Dr. Pritam K. Gujarati, University of Technology and Applied Sciences, Nizwa, Oman
- Dr. Rajaram Shinde, Rajarambapu Institute of Technology, Maharashtra, India
- Dr. Rajneesh Kumar, Birla Institute of Technology and Science, Pilani, Pilani, India
- Dr. S. D. Yadav, Rajarambapu Institute of Technology, Rajaramnagar, Maharashtra, India
- Dr. S. K. Tyagi, Center for Energy Studies, Indian Institute of Technology Delhi, India
- Dr. S. R. Kumbhar, Rajarambapu Institute of Technology, Rajaramnagar, Maharashtra, India
- Dr. S. R. Patil, Rajarambapu Institute of Technology, Maharashtra, India
- Dr. Sandeep R. Desai, Rajarambapu Institute of Technology, Maharashtra, India
- Dr. Sayak Banerjee, Indian Institute of Technology, Hyderabad, India
- Dr. Sonal K. Thengane, Indian Institute of Technology Roorkee, India
- Dr. Sudheer Siddapureddy, Indian Institute of Technology Dharwad, Karnataka, India
- Dr. Sulabh Sachan, FET, Bareilly, India
- Dr. Sunil K. Maity, Indian Institute of Technology, Hyderabad, Telangana, India
- Dr. Utkarsh Raj, Electrical Engineering, Motihari College of Engineering, Motihari, India
- Dr. Venkatasailanathan R., Indian Institute of Technology, Bombay, Powai, Mumbai, India
- Dr. Vijay Khawale, Yeshwantrao Chavan College of Engineering, Nagpur, India
- Dr. Yogendra Arya, Maharaja Surajmal Institute of Technology, New Delhi, India

IOP Conf. Series: Earth and Environmental Science 1042 (2022) 011001 doi:10.1088/1755-1315/1042/1/011001

Advisory Committee

- Dr. A. Ramesh, Chennai Institute of Technology, Kundrathur, Chennai, India
- Dr. A. Senthil Kumar, Sanskrithi School of Engineering, Puttaparathi, India
- Dr. C. M. Sedani, TSSM'S Padmabhooshan Vasantdada Patil Institute of Technology, India
- Dr. D.G. Thombare, Rajarambapu Institute of Technology, Maharashtra, India
- Dr. D. Patnaik, Government College of Engineering, Kalahandi, Bhawanipatna, Odisha, India
- Dr. Dola Sanjay S, Ramachandra College of Engineering, Eluru, Andhra Pradesh, India
- Dr. G L Valsala, College of Engineering Muttathara, Vallakadavu, Thiruvananthapuram, India
- Dr. G Srinivasulu, Narayana Engineering College, Nellore –Andhra Pradesh, India
- Dr. Kalyana Chakravarthy, Surya School of Engineering and Technology, Tamilnadu, India
- Dr. S. M. Sawant, Rajarambapu Institute of Technology, Maharashtra, India
- Dr. Sanjay H. Sawant, Sant Gajanan Maharaj College of Engineering, Mahagaon, Kolhapur, India
- Dr Suresh Babu, Sree Buddha College of Engineering, Pattoor, Kerala, India
- Dr. Swami Naidu Gurugubelli, JNTUK University College of Engineering, India

Organizing Committee

Members from Alard College of Engineering and Management, Pune, India

- Mrs. Ankita Rajput
- Mrs. Chetna Baviskar
- Mr. Deepak Malgar
- Mr. Ganesh Shitole
- Mrs. Lata Kotwani
- Mrs. Nivedita K
- Mrs. Pooja Chandgude
- Mrs. Pooja Patil
- Mrs. Rachna Vaidya
- Mrs. Reshma Patil
- Mrs. Zarina Shaikh
- Mr. Yasmin Khan
- Mr. Shivaji Sathe

Members from Accra Technical University, Accra, Ghana

- Dr. Bernard Arthur-Aidoo
- Prof. Edmund Ameko
- Dr. Isaac Asempa
- Dr. Kwabena Glenn Gyimah
- Prof. Mathias B. Michael
- Dr. Moro Adams
- Dr. Nana Yaw Asabere
- Dr. Peter Nyanor
- Dr. Phyllis Naa Yarley Otu
- Dr. Prodeo Y. Agbotui
- Dr. Stephen Bani
- Dr. Tina Dzigbordzi Wemegah

IOP Conf. Series: Earth and Environmental Science 1042 (2022) 011002 doi:10.1088/1755-1315/1042/1/011002

Peer Review Statement

All papers published in this volume have been reviewed through processes administered by the Editors. Reviews were conducted by expert referees to the professional and scientific standards expected of a proceedings journal published by IOP Publishing.

- **Type of peer review:** Single Anonymous
- **Conference submission management system:** EasyChair
- **Number of submissions received:** 63
- **Number of submissions sent for review:** 32
- **Number of submissions accepted:** 19
- **Acceptance Rate (Submissions Accepted / Submissions Received × 100):** 30.2
- **Average number of reviews per paper:** 2
- **Total number of reviewers involved:** 14
- **Contact person for queries:**
  Name: Sandip A. Kale
  Email: sandipkale@technologycentre.co.in
  Affiliation: Technology Research and Innovation Centre, Pune, India

IOP Conf. Series: Earth and Environmental Science 1042 (2022) 012001 doi:10.1088/1755-1315/1042/1/012001

Influence of different biodiesel fuels on the PCCI-DI combustion concept for constant speed genset application

Hemant A Kinikar[1*], **A B Kanasepatil**[1] **and S S Thipse**[2]

[1]Sinhgad College of Engineering, Savitribai Phule Pune University, Pune, India

[2]Automotive Research Association of India (ARAI), Pune, India

*Email: hakinikar@rediffmail.com

Abstract. Homogenous charge compression ignition (HCCI) is a technology for concurrently reducing nitrogen oxides (NOx) and particulate matters (PM) emissions. Since, the pure HCCI lags in control over combustion phasing, Pre-mixed charge compression ignition (PCCI) method is proposed. To achieve partial homogeneity in PCCI combustion, the air is pre-injected with fuel. The main injection is required for the combination to take place. This paper investigates the impact of using the varying percentages and types of bio-diesels for the constant speed genset application engine. The paper first discus the effect of various ratios of diesel and biodiesel mixtures on the performance parameters like engine power, specific fuel consumption (sfc) and the break thermal efficiency. These effects are caused by the combustion happening inside the combustion chamber which can be seen in pressure, temperature and heat release curves. The result of the simulation finally discusses the formation of NOx and PM emissions during the combustion. These results also provide guidelines for actual testing of the engine with bio-diesel.

1. Introduction

The new emission regulations are based on these health consequences due to emissions. After-treatment devices such as Selective Catalytic Reduction (SCR), Diesel Oxidation Catalyst (DOC), and others are used to comply with these emission regulations. The use of after-treatment devices adds to the total cost. In order to lower the overall cost, combustion must be improved. One method for improving combustion is a Homogeneous Charge Compression Ignition (HCCI). Since the pure HCCI is based on the chemical kinetics and time-temperature history for start of combustion. The PCCI concept retains control over the start of combustion while giving benefits of HCCI. In this paper, the simulation is used to examine the different bio-diesels as an alternative fuel for engine combustion. The engine's performance and emissions will alter as the fuel is changed. This research examines this modification in the context of a constant-speed genset engine.

2. Methodology

Before beginning, the PCCI concept must be simulated and analyzed on a diesel genset engine. The background and context of the simulation study must be understood. The two points listed below provide approaches to comprehending the situation.

- Need of simulation
- Fuel variations used for simulation

2.1. *Need of simulation*

Knocking at high loads and misfire at low loads limit HCCI [1]. As a result, PCCI – DI is preferred to in place of pure HCCI. The engine operates in PCCI mode at lower loads in PCCI – DI combustion. During larger loads, it switches to DI mode. Because of this dual function, the engine must be optimized for both DI and PCCI operations. Before the engine is really tested with the different bio-diesels, the simulation is the best technique to analyze the PCCI-DI combustion. Various researchers have conducted the actual experiments with the [2–6] different biodiesels with diesel on a PCCI engine. However, this paper evaluates the effect of using bio-diesel through simulation.

2.2. *Finalization of simulation software*

Significant numbers of simulation software are already available in the market. The survey of the simulation software for engine combustion reveals the following list of software to be used in PCCI study.

- Diesel RK software [7]
- Converge [8]
- Star CCM+ [9]
- KIVA [10], [11]
- Ricardo Wave [12]
- AVL Boost [12]
- AVL fire [13]
- GT Power [14]
- MATLAB [15]

Based on the market survey and the availability of software, Diesel - RK software is used for this analysis [7].

2.3. *Fuel variations for used in the simulation*

The following are the different bio-diesel used in simulation.

- Diesel
- B5 SME (Soy Methyl Ester)
- B20 SME (Soy Methyl Ester)
- B40 SME (Soy Methyl Ester)
- B100 SME (Soy Methyl Ester)
- B100 RME (Rapeseed Methyl Ester)

This simulation study makes use of a change in fuel and its effect on the diesel engine's current configuration. This will show how utilizing bio-fuel instead of pure diesel affects the performance and emissions. The engine configuration for the use of PCCI-DI mode of operation for the diesel fuel is shown in Table 1.

2.4. *Selection of modes for PCCI operation*

The modes of operation refer to the precise conditions of engine operation in which emission parameters are measured. The 5 Mode cycle is used to simulate the genset engine. The power produced in that mode, as well as the brake mean effective pressure (BMEP) readings, is used to choose the mode for PCCI deployment. According to the literature review available [16,17], mode 4 and mode 5 are used to execute the PCCI approach, as shown in Table 1. Hence, the range of 1 bar to 6.5 bar BMEP is chosen based on the literature available [16,17].

The severity attributed to a certain mode during the emission certification test cycle is shown by the weightage column in Table 1. As a result, PCCI assessment is limited to the 4th and 5th modes. These two modes affect the engine emissions on 40% of the total engine mass emission test cycle.

Table 1. PCCI mode selection in diesel genset engine

Mode	Speed RPM	Torque N-m	Power kW	BMEP bar	Mode of operation	Weightage %
1	1500	400	62.9	13.6	DI Mode	5
2	1500	300	47.1	10.2	DI Mode	25
3	1500	200	31.4	6.8	DI Mode	30
4	1500	100	15.7	3.4	PCCI Mode	30
5	1500	40	6.3	1.4	PCCI Mode	10

3. Simulation

Different combustion parameters must be defined in order to simulate engine combustion. The simulation parameters range from general specifications to combustion geometry. The several parameters used as input to the simulation are discussed and finalized in this section. These parameters are divided into the following subgroups.

- General specifications
- Combustion geometry specifications
- Fueling details
- Other parameters

3.1. General specifications

Table 2 contains general information and data. The bore, stroke, and connecting rod length are the essential parameters that provide information about the engine. The 1500 RPM genset power is specified as it us used for genset application.

Table 2. Basic engine details

Physical Parameters	Value	Unit	Remarks
Bore	95	mm	Cylinder bore diameter
Stroke	130	mm	Stroke of piston
Connecting rod length	230	mm	Distance between the small & big end center
Crank offset	65	mm	
Swept volume per cylinder	921.6	CC	
Compression ratio	17.5	Ratio	
Crank speed	1500	RPM	
Rated power	63	kW	

3.2. Valve Lift Details

The maximum valve lift specifications of the inlet valve and the exhaust valve are as follows.

- Inlet 11.5 (mm)
- Exhaust 11.5 (mm)

Because the valves are opened mechanically by the camshaft, tappet, and rocker mechanism, the engine uses fixed valve opening and closing timings independent of load. Figure 1 depicts the exact details of opening and closing the intake and output valves.

3.3. Fueling Specifications

The main fueling inputs are described in Table 3 and the additional inputs are listed in the operating mode Table 4. In the Diesel-RK software, each mode's simulation run is done separately. The five modes of genset emission testing points in DI mode are used initially in this investigation. Then it looks at the impact of using various fuels instead of diesel. Since the purpose of the paper is to evaluate the effect bio-fuels, the EGR is not used during the assessment. In view of the same, the EGR ratio is not specified in the simulation.

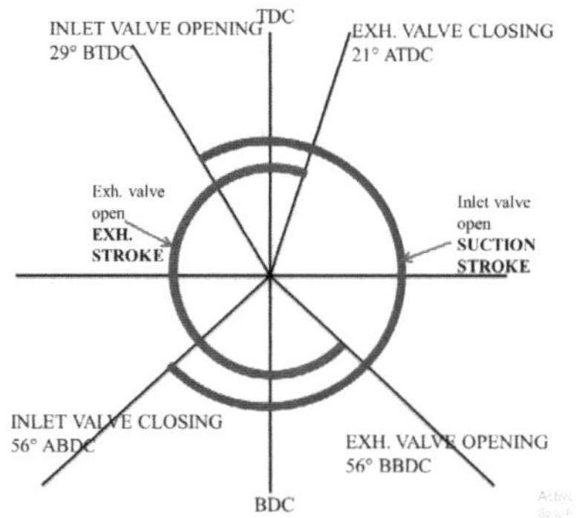

Figure 1. Engine valve timing diagram

Table 3. Base engine fuel injection timing and bifurcation data at 1500 RPM

Mode	Power	Brake Torque	Pre Injection	Pre Injection	Main Injection	Main Injection	Fuel injected	Inj. Pr.
	kW	N-m	%	°bTDC	%	°bTDC	g/stroke	bar
1	62.9	400	2.2	17	97.8	1.5	0.0681	1063
2	47.1	300	2.9	17	97.1	1.5	0.0522	952
3	31.4	200	3.9	17	96.1	1.5	0.0380	918
4	15.7	100	6.7	17	93.3	1	0.0226	877
5	6.3	40	11.2	17	88.8	0.5	0.0134	800

Table 4. The performance values of the diesel genset engine through simulation

Parameter	Unit	Mode 1	Mode 2	Mode 3	Mode 4	Mode 5
Combustion type	DI/PCCI	DI	DI	DI	PCCI	PCCI
Engine Power	kW	62.9	47.1	31.6	16.0	6.4
Indicated Efficiency	bar	0.47	0.48	0.47	0.46	0.43
Efficiency of engine	Fraction	0.44	0.43	0.40	0.34	0.23
Max. Cylinder Pr.	bar	138	120	92	80	66
Mass of Fuel	g/stroke	0.068	0.052	0.037	0.0218	0.013
SFC	kg/kWhr	0.195	0.199	0.211	0.246	0.364
Start of Combustion	°bTDC	10.43	9.94	9.35	11.12	6.58
Ignition Delay Period	°	6.57	7.06	7.65	8.88	7.42
Intake Airflow	kg/s	0.1031	0.0901	0.0761	0.0762	0.0640
Lambda	Ratio	2.09	2.39	2.84	4.82	6.79
Fraction of wet NOx	PPM	1946	1699.3	1246.4	389.79	168.38
Specific PM emission	g/kWh	0.072	0.102	0.145	0.153	0.443

The simulation engine details have previously been optimized for work with PCCI-DI modes. The other simulation inputs are not discussed in this paper. Table 4 shows the fuelling and other major parameters in detail.

3.4. Simulation Assessment Process

Before the start of the simulation, process of simulation needs to be finalized. This is done in the following sections. The existing diesel engine details will be used as the simulation's baseline data. This baseline will serve as a basis of comparison for evaluating the performance of various biofuels. The only change is in the chemistry of the fuel injected to simulate the various bio-fuels utilized without requiring any modifications to the diesel engine.

4. Results and Discussions

The results of the simulations are divided in terms of performance and emissions. These are discussed in the following sections in tabular and graphical format.

4.1. Simulation results – Performance comparison

First, a simulation model analysis of the existing engine data in DI mode is completed. This primary data analysis serves as a starting point for comparing modifications. The variation in engine power due to fuel change is also shown in
Figure **2**.

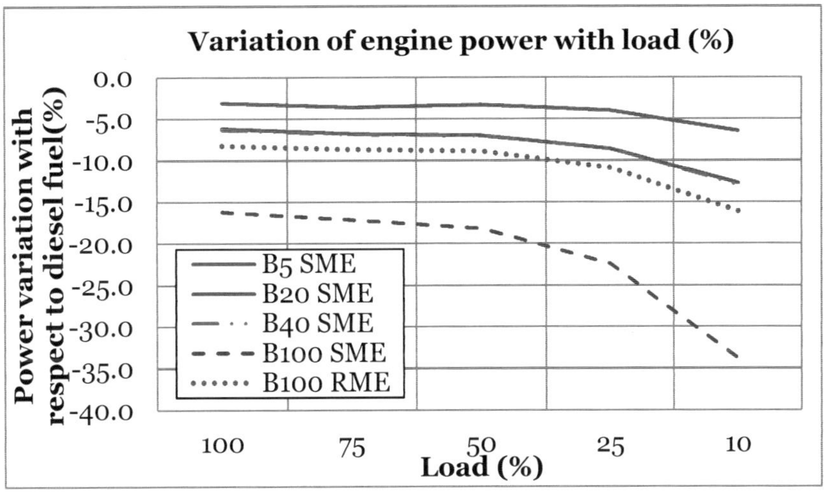

Figure 2. Variation of engine power with load (%)

Figure 3 shows that the diesel-powered engine consumes less fuel than the biodiesel-powered engine. This is because diesel has a high calorific value and has a lower heat of vaporization. As shown in Figure 3, the SFC increases (deteriorates) as the percentage of biofuel in the fuel increases. At higher loads, this effect is comparatively smaller. However, as the load decreases and the bio- fuel concentration in the fuel increases, this effect becomes savior. At full load, the best SFC is observed with diesel-fueled engines. If the diesel engine performance is taken as a base line, there is a significant increase/deterioration in sfc with B100 SME as shown in Figure 3.

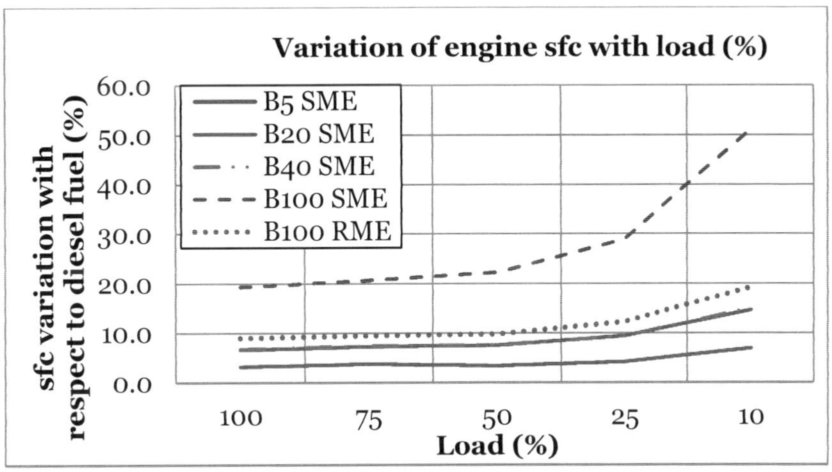

Figure 3. Variation of engine SFC with load (%)

Figure **4** show the variation in brake thermal efficiency. The use of bio-fuels results in a loss in fuel efficiency. This is an indication that using bio-fuel, adversely affects the engine's efficiency. It also shows that at rated load, the reduction is negligible, but at 10% load, the reduction is 22.5% using B100 SME fuel. In comparison to other fuels, the B20 SME's results demonstrate a lower rate of deterioration. With the high load and the lower percentage of the biofuel the effect is not significant at higher load. Engine efficiency is reduced when the load is reduced or when the percentage of bio-fuel is increased.

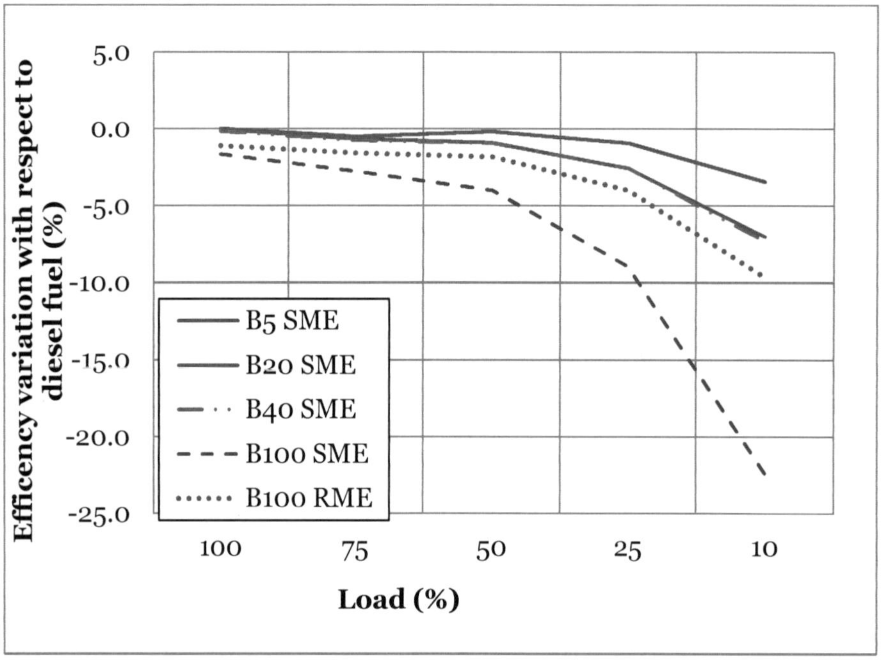

Figure 4. Variation of engine efficiency with load (%)

4.2. Simulation results – Emission comparison

Figure 5 shows the relationship between brake power and nitrogen oxides (NOx). Because of the high temperature inside the combustion chamber, NOx has been produced. Figure 5 shows the NOx variation at various loads. At higher and medium loads, bio-fuels provide reduced/improved NOx performance. As the load decreases, however, the trend reverses, and biofuels emit more NOx than diesel engines. At lower loads, the B100 RME produces the maximum NOx. At lesser loads, the other fuels only show a moderate increase in NOx.

The biggest disadvantage of the CI mode engine is particulate pollution. It is created as a result of incomplete combustion. The trend in PM is seen in Figure 6. Only under higher loads and with a lower percentage of bio-fuel in the fuel can the cause increased PM emissions. PM emissions drop as the percentage of biofuel increases and the load decreases. As illustrated in Figure 6 using pure bio-diesel results in significant reductions of PM at all loads. Because of the presence of oxygen within the biofuel, this is the case.

4.3. Simulation results –Tabular representation

This section discusses the two main parameters of the combustion chamber as follows. Maximum cylinder pressure (refer Table 5) Maximum combustion temperature (refer Table 6). As shown in Table 5, regardless of the fuel used, the maximum cylinder pressure for a particular load remains fairly constant. This indicates that a change of fuel has no substantial influence on peak cylinder pressure.

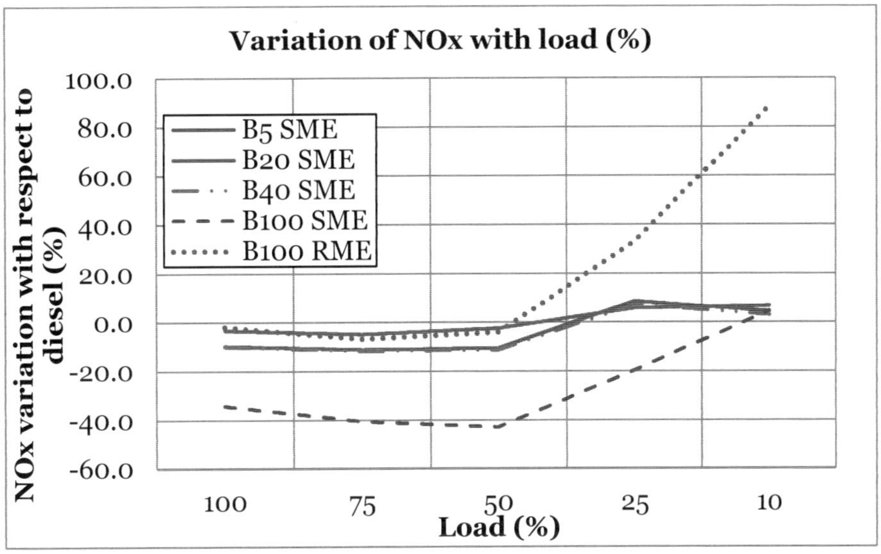

Figure 5. Variation of NOx with load (%)

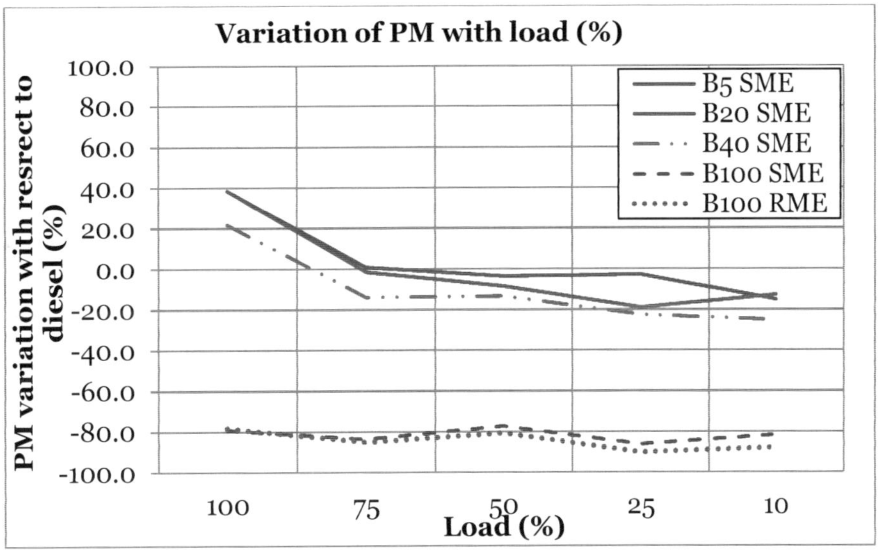

Figure 6. Variation of PM with load (%)

Table 5. Maximum cylinder pressure

Mode	Speed RPM	Load %	Cylinder pressure – Maximum (Bar) Diesel	B5 SME	B20 SME	B40 SME	B100 SME	B100 RME
1	1500	100	138.2	139.6	139.3	139.2	132.4	134.5
2	1500	75	119.9	120.4	119.4	119.8	113.7	115.9
3	1500	50	92.0	96.6	95.3	96.2	91.4	93.2
4	1500	25	79.9	83.6	82.7	83.2	82.9	83.8
5	1500	10	66.5	67.1	66.6	66.9	68.3	69.5

When bio-fuel is used, the maximum cylinder temperature is lower for higher and moderate loads on the engine. Table 6 and Table 7 show this. The maximum temperature is reduced as the percentage of biofuel increases. However, with lower loads and with a lower percentage of bio-fuel in the fuel, there is a slight increase in-cylinder temperature.

Table 6. Maximum combustion temperature

Mode	Speed RPM	Load %	Cylinder pressure – Maximum (°K) Diesel	B5 SME	B20 SME	B40 SME	B100 SME	B100 RME
1	1500	100	1822	1783	1803	1781	1677	1736
2	1500	75	1752	1702	1720	1698	1600	1659
3	1500	50	1618	1594	1615	1592	1494	1545
4	1500	25	1297	1313	1311	1310	1269	1303
5	1500	10	1141	1158	1149	1155	1161	1189

Table 7. Maximum combustion temperature w.r.t diesel in percentage

Mode	Speed	Load	% Variation B5 SME	B20 SME	B40 SME	B100 SME	B100 RME
1	1500	100	-2.1	-1.0	-2.2	-8.0	-4.7
2	1500	75	-2.8	-1.8	-3.0	-8.6	-5.3
3	1500	50	-1.5	-0.2	-1.6	-7.7	-4.5
4	1500	25	1.2	1.1	1.0	-2.2	0.4
5	1500	10	1.5	0.7	1.2	1.8	4.3

4.4. Simulation results – Graphical representation

The use of graphical visualization is preferred for simplicity of understanding. The simulation output is depicted graphically in the diagram below. Because all of the modes' details cannot be compared. Only the variation at 100% load will be examined here. This will allow for improved parameter visualization.

- Cylinder pressure against crank angle graph for 100% load (P-theta diagram) (refer Figure 7)
- Combustion temperature against crank angle graph for 100% load (refer
- Figure **8**)
- NOx (PPM) against crank angle graph for 100% load (refer Figure 9)
- Soot against crank angle graph for 100% load (refer
- Figure **10**)
- The heat release rate against crank angle graph for 100% load (refer Figure 11)
- Heat release fraction against crank angle graph for 100% load (refer Figure 12)

Figure 7 shows the cylinder pressure curves obtained from the simulation analysis. Even when the fuel is changed, the peak pressure achieved practically stays in the same range, according to a detailed analysis of the pressure traces.

Figure **8** compares the cylinder temperatures in the PCCI and DI modes. Figure 7 shows a pressure trend that differs from the temperature trend presented. The change can be seen in the numerically specified peak cylinder temperature in Table 6.

Figure 9 shows a detailed analysis of NOx generation in the combustion chamber. In Figure 9 a detailed analysis of NOx generation in the combustion chamber. In comparison to other fuels, the B100 SME fuel produces the lowest NOx emissions. As seen in

Figure **8**, this occurrence can be attributable to the lower cylinder temperature created inside the combustion chamber.

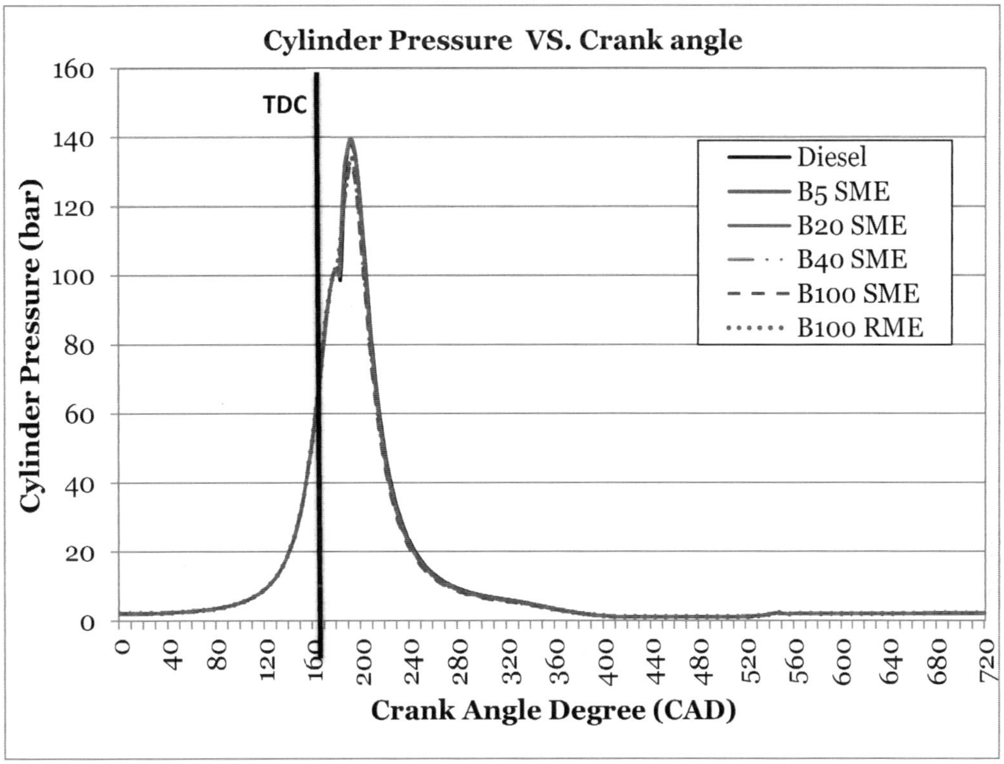

Figure 7. Cylinder Pressure against crank angle graph (P-theta diagram)

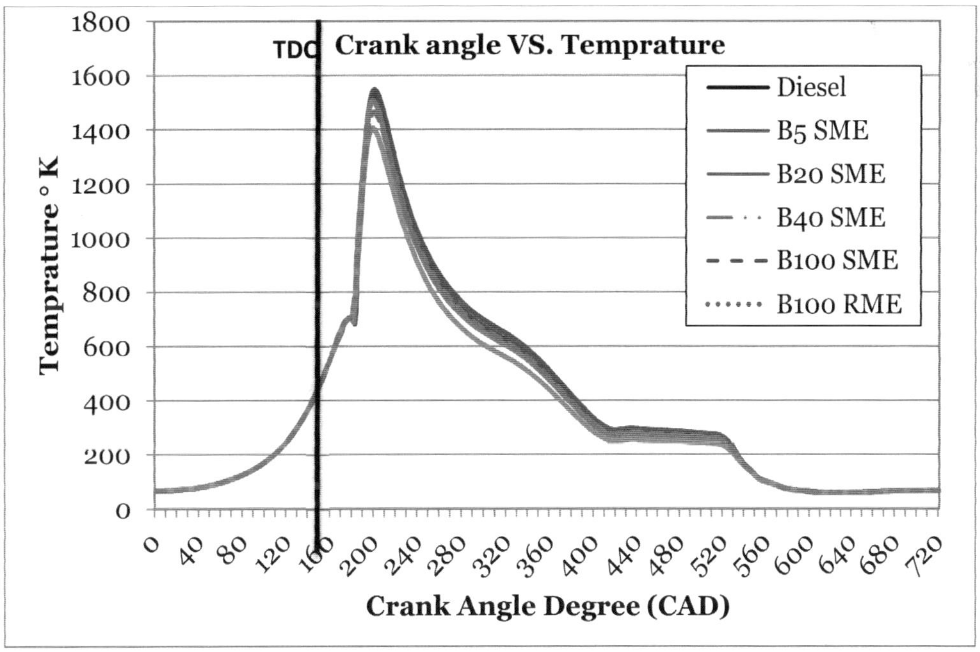

Figure 8. Combustion temperature against crank angle graph

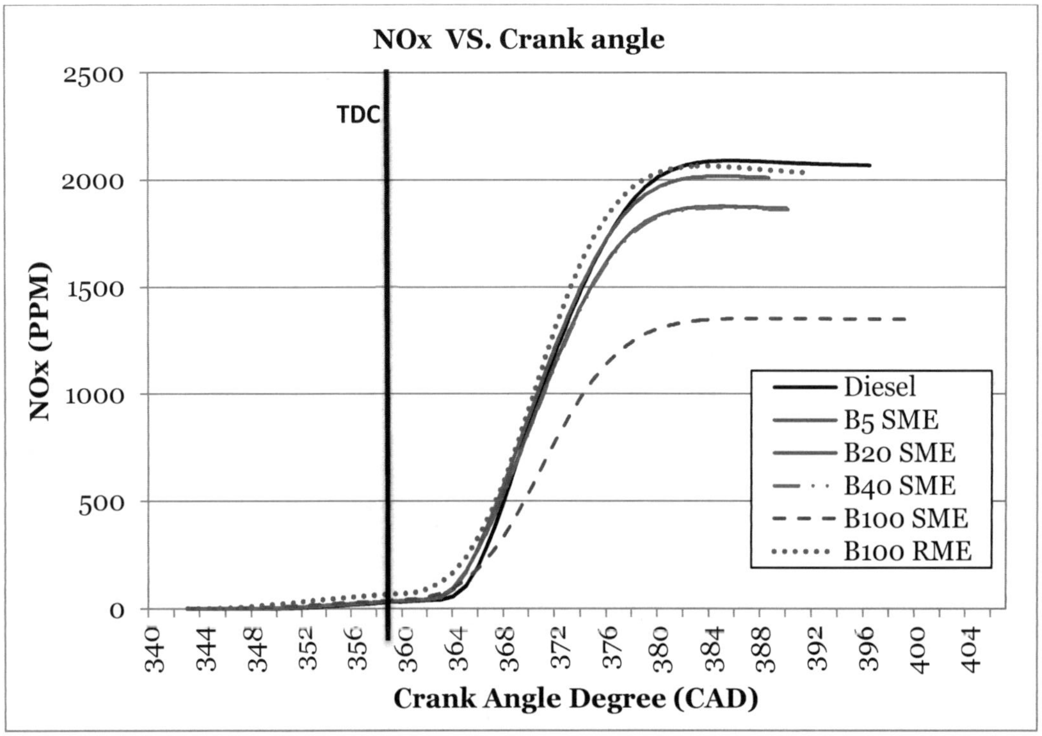

Figure 9. NOx (PPM) against crank angle graph

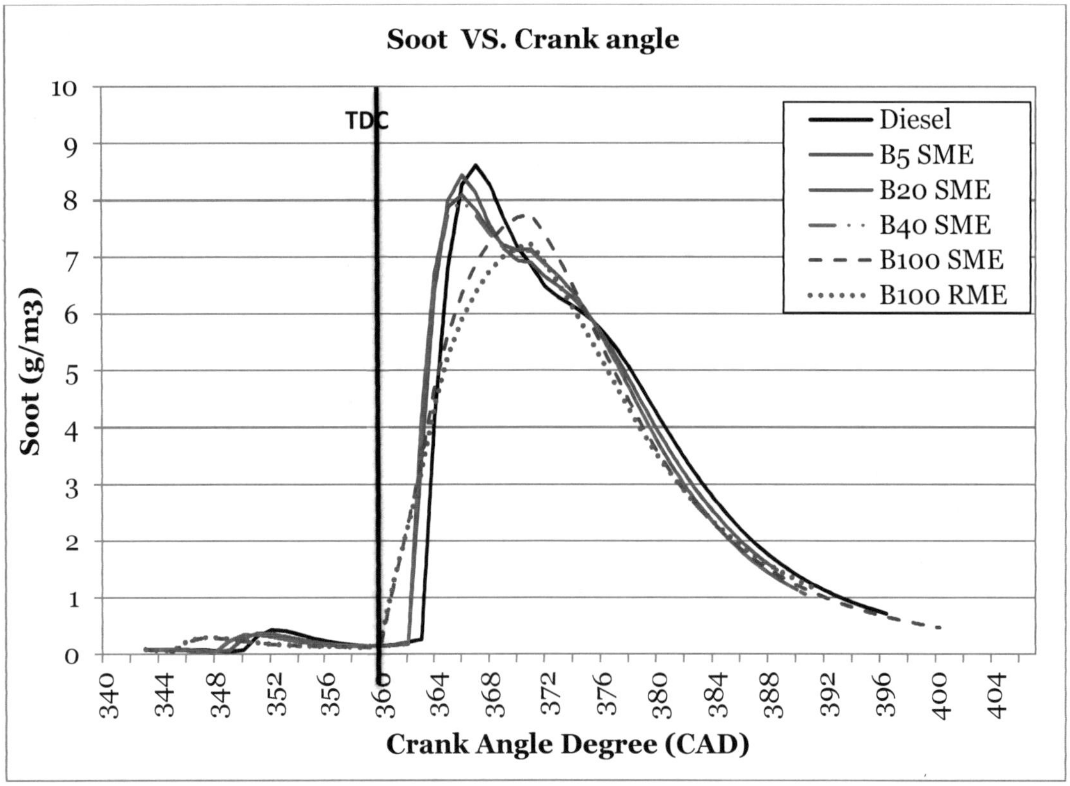

Figure 10. Soot against crank angle graph

Unburned fuel causes soot to collect inside the combustion chamber.

Figure **10** shows the soot production in detail as a function of crank angle. This formation rate of soot is steep for the diesel and other blends while the soot formation rate is slow for the B100 SME and B100 RME. As indicated in Figure 11, both of these factors result in less soot being converted to particle emissions.

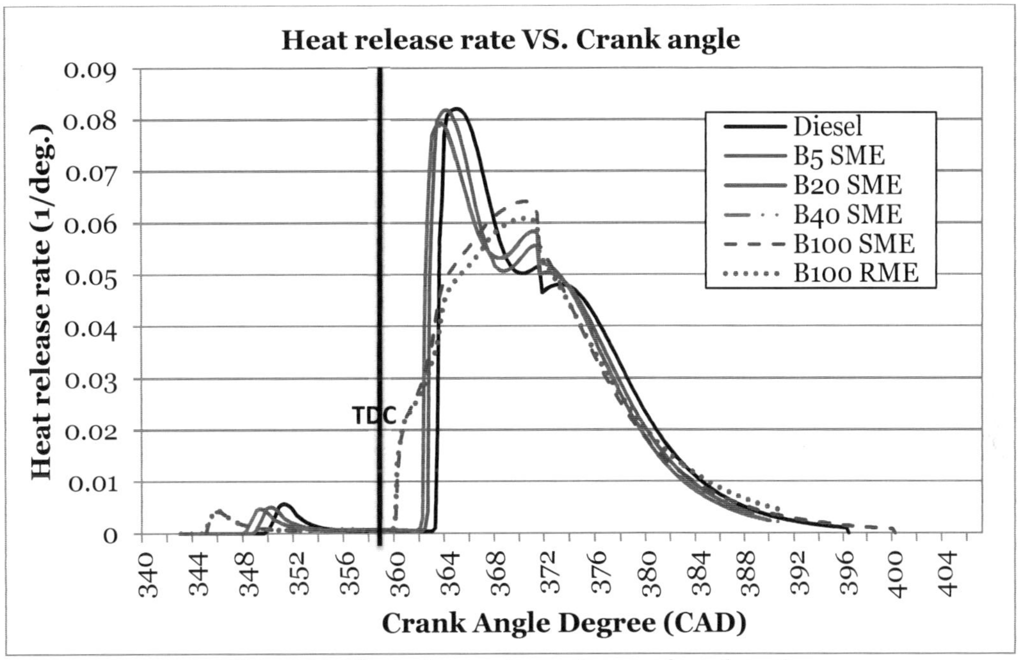

Figure 11. Heat release rate per degree of crank angle

Figure 11 and Figure 12 show the heat release happening in the combustion chamber. The heat release rate diagram Figure 12 is only as additional information.

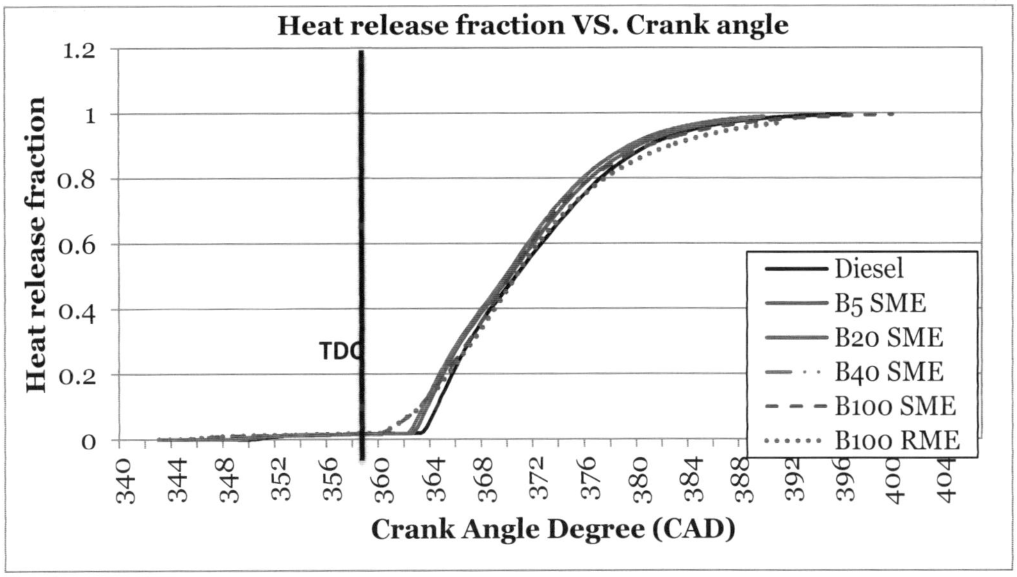

Figure 12. Heat release fraction against crank angle graph

5. Conclusions

The conclusions of the simulation can be summarized in the following points.

- Use of biofuel reduces the engine power as the calorific value of bio-fuels are lesser than the diesel fuel.
- For the same quantity of fuel injected, the reduction in load is based on the % and type of biodiesel. This reduction in load aggravates as the load requirement drops.
- The NOx emission reduces with use of biofuel at moderate and higher loads. However, it shows an opposite trend at lower loads. B100 RME gives significantly high NOx at lower loads.
- PM emissions are significantly lower when 100% biofuel is used. However, for the fuels with lesser amount of bio-diesel. The PM is higher at the rated load and reduces when the load is reduced.
- As a way forward, actual engine testing can be conducted to assess the simulation results.

References

[1] Juttu S, Thipse S S, Marathe N V., Babu G M K and Andersson A I 2009 CFD Study of Combustion Chambers for Lower Engine Exhaust Emissions from Diesel Engines Operated in HCCI and Conventional Diesel Mode *SAE Technical Papers* vol 2009-Janua

[2] Sendilvelan S and Bhaskar K 2016 Experimental analysis of pilot injection-assisted premixed charge compression ignition to reduce emissions with jatropha oil methyl ester in a diesel engine *Int. J. Energy a Clean Environ.* **17** 67–80

[3] Srihari S and Thirumalini S 2017 Investigation on Reduction of Emission in PCCI-DI Engine with Biofuel Blends *Renew. Energy*

[4] Srihari S, Thirumalini S and Prashanth K 2017 An Experimental Study on the Performance and Emission Characteristics of PCCI-DI Engine Fuelled with Diethyl ether-Biodiesel-Diesel Blends *Renew. Energy*

[5] Natarajan S, Kumar M A and Sundareswaran A U M 2017 Computational Analysis of an Early Direct Injected HCCI Engine Using Bio Ethanol and Diesel Blends as Fuel *Energy Procedia* **105** 350–7

[6] Kumar P and Rehman A 2016 Bio-diesel in homogeneous charge compression ignition (HCCI) combustion *Renew. Sustain. Energy Rev.* **56** 536–50

[7] Kuleshov A and Grekhov L 2013 Multidimensional optimization of di diesel engine process using multi-zone fuel spray combustion model and detailed chemistry NOx formation model *SAE Technical Papers*

[8] An Y, Jaasim M, Raman V, Hernández Pérez F E, Sim J, Chang J, Im H G and Johansson B 2018 Homogeneous charge compression ignition (HCCI) and partially premixed combustion (PPC) in compression ignition engine with low octane gasoline *Energy* **158** 181–91

[9] Latheesh V M, Parthasarathy P, Baskaran V and Karthikeyan S 2018 Design and CFD analysis of intake port and exhaust port for a 4 valve cylinder head engine *IOP Conference Series: Materials Science and Engineering* vol 310

[10] Kong S C, Marriott C D, Reitz R D and Christensen M 2001 Modeling and experiments of HCCI engine combustion using detailed chemical kinetics with multidimensional CFD *SAE Technical Papers*

[11] Zehni A and Saray R K 2018 Comparison of late PCCI combustion, performance and emissions of diesel engine for B20 and B100 fuels by KIVA-CHEMKIN coupling *Renew. Energy*

[12] Alqahtani A, Shokrollahihassanbarough F and Wyszynski M 2015 Thermodynamic simulation comparison of AVL BOOST and Ricardo WAVE for HCCI and SI engines optimisation *Combust. Engines*

[13] Musu E, Gentili R and Reitz R D 2009 Homogeneous Charge Progressive Combustion (HCPC): CFD study of an innovative diesel HCCI concept *SAE Technical Papers*

[14] Mamalis S, Nair V, Andruskiewicz P, Assanis D, Babajimopoulos A, Wermuth N and Najt P

2010 Comparison of different boosting strategies for homogeneous charge compression ignition engines - a modeling study *SAE Int. J. Engines* **3** 296–308

[15] Keshavarz M and Jazayeri S A 2006 Performance of Homogenous Charge Compression Ignition (HCCI) engine with premixed methane/air supported by DME for electrical power generation application *American Society of Mechanical Engineers, Internal Combustion Engine Division (Publication) ICE*

[16] Gowthaman S and Sathiyagnanam A P 2018 Analysis the optimum inlet air temperature for controlling homogeneous charge compression ignition (HCCI) engine *Alexandria Eng. J.* **57** 2209–14

[17] Juttu S, Mishra P, Thipse S S, Marathe N V and Babu M K G 2011 Combined PCCI-DI Combustion to Meet EURO-IV Norms on LCV Engine - Experimental and Visulisation Study *SIAT 2011* 1–10

Investigation of Mahua, Pongamia and waste cooking oil blends as biodiesel properties

Malati Chanagond[1*], Prashant Puthani[2], Ramesh Jeeragal[2], Suresh Doddi[2]

[1] Department of Chemistry, BLDEA's S B Arts and K C P Science College, Vijayapur-586103, Karnataka, India
[2] Department of Mechanical Engineering, BLDEA's V. P. Dr. P. G. Halakatti College of Engineering and Technology, Vijayapur-586103, Karnataka, India

[*]E-mail: chanagondmalati@gmail.com

Abstract. Looking at the present scenario of world's energy consumption and its dependency on fossil fuels, biodiesel has become a ray of hope as it is environmental friendly fuel derived from plant sources. In that context the present paper reports the study of physical properties of Mahua, Pongamia and Waste cooking oil blend for usage as biodiesels in engines. Each biodiesel has its own specific properties which need to be tested before running test on a engine. The physical properties such as kinematic viscosity, flash point, fire point, density, heating values and acid values are measured for MPW (Mahua 50%, Pongamia 30%, Waste Cooking Oil 20%), PWM (Pongamia 50%, Waste Cooking Oil 30%, Mahua 20%) and WMP (Waste Cooking Oil 50%, Mahua 30%, Pongamia 20%) compositions prepared on volume basis. This study discovered that WMP biodiesel physical properties match nearly to that of diesel fuel. It is the best alternative with moderate flash point temperature, fire point temperature, better thermal efficiency, density and acid value in comparison with MPW and PWM biodiesels.

1. Introduction

Increase in the energy demand of fossil fuels for transportation also industrial sector has led to environmental pollution with rise in global warming year by year. To curb on such issues researchers have been motivated to opt for alternative fuels which can meet the properties of fossil fuels, decrease pollution, reduction in dependency of fossil fuels as well as improvement in country's economy. Many plant oil derivatives in the nature have some desirable properties of the fossil fuels which need to be estimated before using in the engines. Non edible oils are highly preferred as they cannot be used for cooking or any other purposes, therefore extraction of oil from these non edible seeds can be a substitute to the fossil fuels. From researches it is found that biodiesel physical properties nearly match with diesel. Depending on their properties biodiesel can be used with less or without any modifications in the engine.

Dario Alviso et al. [1] reported the refractive index, kinematic viscosity and flash point of soyabean, peanut, canola, olive , corn, almond and grape blends for the determination, it is found that R^2 values are larger than those found using simple correlations. Cijil Biju John et al. [2] investigated kinematic viscosity, density and cetane number properties of hemp biodiesel to be 3.91cSt, 876kg/m^3 and 50. Arun et al. [3] assessed the physiochemical properties of yellow oleander, simarouba glauca and dairy scum to conclude that there was a significant difference between the physiochemical properties of biodiesel and petroleum diesel. D.Ravichandra et al. [4] studied the properties of neem, cotton seed and sesame oils and concluded that the basic properties of these oils nearly match with diesel. Sani et al. [5] conducted properties test on citrullus lanataus biodiesel blends of B5, B10, B15, B20 and B25 fuels. He concluded that B20 blend calorific value is below than that of diesel, flash point and fire point temperature is higher than that of diesel also fuel can be safely stored. N. Acharya et al. [6] concluded that the biodiesel properties of mahua are within the specifications standards with 30% blend preferred considering cold filter plugging point. Mamdouh T et al. [7] investigated properties of jojoba biodiesel and concluded that it had high flash point temperature, less ash and lower sulfur contents to that of diesel. Gerald Kafuku et al. [8] conducted test on flow properties of biodiesel. He reported that there was a reduction in cloud and pour point values when compared with neat diesel. F. Aydin et al. [9] concluded that methyl ester prepared from corn oil had density and kinematic viscosity properties as per ASTM standards. Osmano Souza Valente et al. [10] reported that maximum biodiesel concentration in the diesel meets the engine characteristics. Roshan Raman et al. [11] concluded that mahua biodiesel with di-ethyl ether improved engine performance and emissions. Harisha P et al.[12] conducted experiment by blending animal fat oil biodiesel, Pongamia biodiesel and waste cooking oil biodiesels in various proportions and reported that the physical properties match nearly with ASTM standards of diesel. Pongamia pinnata is the scientific name of Karanja plant. It is mainly found in subcontinent parts of Asia. This tree is mainly seen on the river banks, coasts which makes it favourable for humid conditions also. Pongamia seeds can yield 30%-40% of oil when compared to other non edible oil seeds. Using Pongamia oil directly in the engine may cause damage to the engine as well as fuel lines may also get choked in winter days because of higher viscosity. Therefore it is converted into biodiesel then blended with diesel in different proportions before conducting engine test. Performance and emission test conducted on engine using Pongamia biodiesel developed good results which matches nearly to that of diesel [13]. Mahua is a plant which belongs to sapotaceae family. It is found in tropical regions of northern part of India. Height is around 20 meters and grows very easily. Large quantity of mahua oil is available and it may be used as a substitute for diesel fuel. As it can be grown in rural and domestic regions very easily, famers can commercialise it as a source of revenue [14]. Oil which cannot be used after cooking is considered to be waste cooking oil. The waste cooking oil is collected from hotels, caterers which is then sent for transesterification process. It is then mixed with diesel by volume and used in engine to check the performance and emission levels [15]. From the literature survey it is observed that very scarce experimental work has been carried on combination of blends of biodiesels in engines. The novelty of this work is to determine the properties of combinations of biodiesel blends and compare with diesel fuel. Also the potentialities of these biodiesels are better, as they are easily available, low cost and a source of employment for rural places.

2. Material and Methods

Pongamia Oil is converted into biodiesel by transesterification method. Here Pongamia oil is heated along with ethanol and sodium hydroxide for 3 to 4 hours maintaining uniform temperature using magnetic stirrer. Then water washing method is carried out so that the Pongamia biodiesel can be collected. Figure 1 shows the Pongamia biodiesel extraction process. Similarly Mahua and Waste cooking oil biodiesels are also prepared.

The main goal of this paper is to opt a biodiesel blend which meets the physical properties of diesel fuel. Figure 2 shows the samples prepared for study. Pongamia biodiesel, waste cooking oil biodiesel and mahua biodiesel are mixed in various proportions to test their properties and compare with diesel fuel. Table 1 represents different biodiesel blends prepared on volume basis.

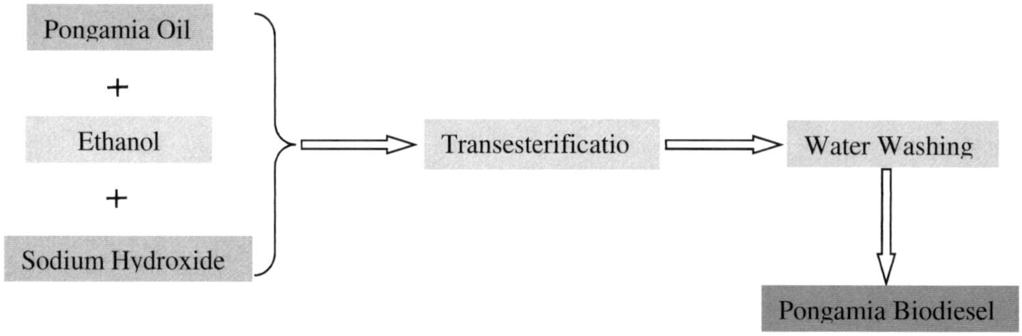

Figure 1. Extraction of Biodiesel

Figure 2. Samples of WMP, PWM, MPW biodiesels

Table 1. Proportions of Different Biodiesels

Sl.No	Proportions (on volume basis)	Name of Mixed Biodiesel		
		MPW	PWM	WMP
1	50%	Mahua	Pongamia	Waste Cooking Oil
2	30%	Pongamia	Waste Cooking Oil	Mahua
3	20%	Waste Cooking Oil	Mahua	Pongamia

3. Results and Discussion

3.1. Flash and Fire Point Estimation Procedure

A Pensky Martens's apparatus, is as shown in figure 3. It is used to calculate the flash point and fire point values of MPW, PWM and WMP biodiesels. Initially first sample of biodiesel i.e. MPW is filled in to the cup until it reaches the marking level. A thermometer & stirrer are placed on the top of a cup

IOP Conf. Series: Earth and Environmental Science 1042 (2022) 012002 doi:10.1088/1755-1315/1042/1/012002

containing biodiesel. Thermometer records the temperature while stirrer maintains uniform temperature of the MPW biodiesel. Heat is supplied to the sample continuously. Once the temperature of the MPW biodiesel reaches 60°C, thereafter for every 3°C-5°C temperature of MPW biodiesel is checked with fire until bluish color flame appears on top of the cup for about 2 seconds to 5 seconds. This value of temperature is known as the fire point. On further heating MPW biodiesel in liquid form, it vaporises to form an ignitable mixture with air to initiate ignition. Now, this temperature of MPW biodiesel where ignition begins is recorded as flash point. The entire procedure is repeated to estimate the flash point and fire point temperatures for PWM and WMP biodiesels using Pensky Marten's apparatus. From figure 4 it can be seen that the flash point and fire point of WMP biodiesel is 64.45% and 63.2% more than that of diesel when compared with other biodiesel combinations. This is due to low volatility, poor vaporization and lesser tendencies to ignite the mixture when compared with diesel.

Figure 3. Pensky Marten's apparataus

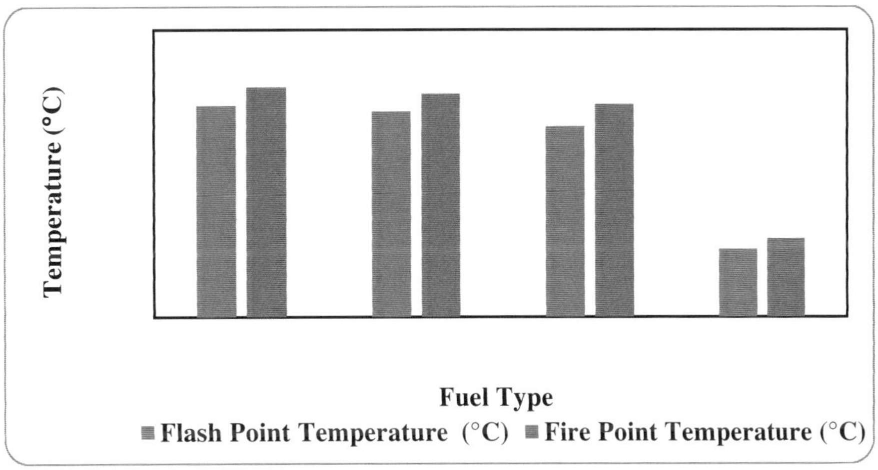

Figure 4. Influence of flash point and fire point for different biodiesel proportions

3.2. Kinematic Viscosity Evaluation Procedure

Viscosity is a physical property which offers resistance to the flow of fluid. It is a function of temperature, which lowers with increase in temperature. Kinematic viscosity is an important property for biodiesels. It is calculated using redwood viscometer. Procedure for conducting test is explained:

initially pour the MPW biodiesel in to a cylindrical orifice cup fitted with a ball valve until it touches the specified mark inside the cylinder. Place a 50ml flask at the bottom for collection of MPW biodiesel as it flows through the orifice. Fill the water bath with water up to a specified mark. Place a thermometer in the cylindrical cup containing MPW biodiesel to record the temperature of biodiesel. A stirrer is placed in water bath maintains uniform temperature of the biodiesels. Temperature of the water bath is controlled with the help of heater which supplies heat to the MPW biodiesel present in the cylindrical cup. Water is stirred continuously to maintain uniform temperature. At 40°C temperature of MPW biodiesel, the ball valve is lifted and MPW biodiesel flows through the orifice in to a 50ml flask. Record the time needed for collection of 50ml biodiesel in a flask. Repeat the procedure for PWM and WMP biodiesels using redwood viscometer as shown in figure 5. Kinematic viscosity values at 40°C for MPW, PWM and WMP is higher when compared with diesel, as viscosity is directly related to chemical structure. For MPW biodiesel the kinematic viscosity is 59.15% more than that of diesel when compared with other two combinations as shown in figure 6. This is due to increase in chain length and decrease in the unsaturation in MPW biodiesel.

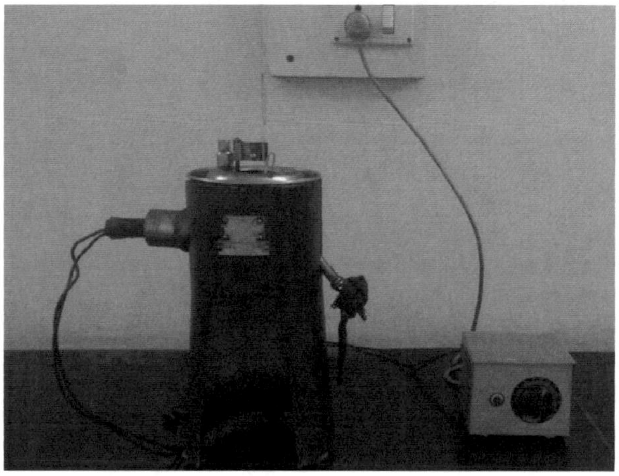

Figure 5. Redwood Viscometer

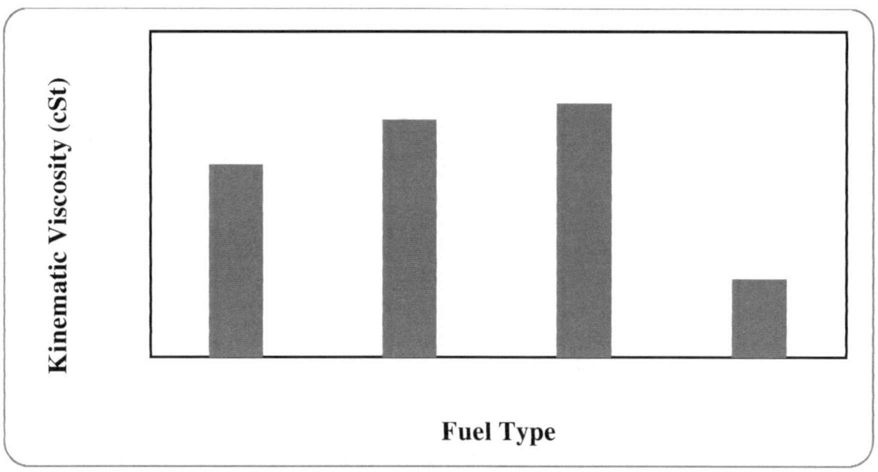

Figure 6. Influence of kinematic viscosity for different biodiesel proportions

3.3. Density Evaluation Procedure
The ratio of mass to volume of a fuel is defined as density. Here the mass of MPW biodiesel is measured for a volume of 50ml at a temperature of 40°C. Density of MPW, PWM and WMP biodiesels are comparable with diesel as shown in figure 7.

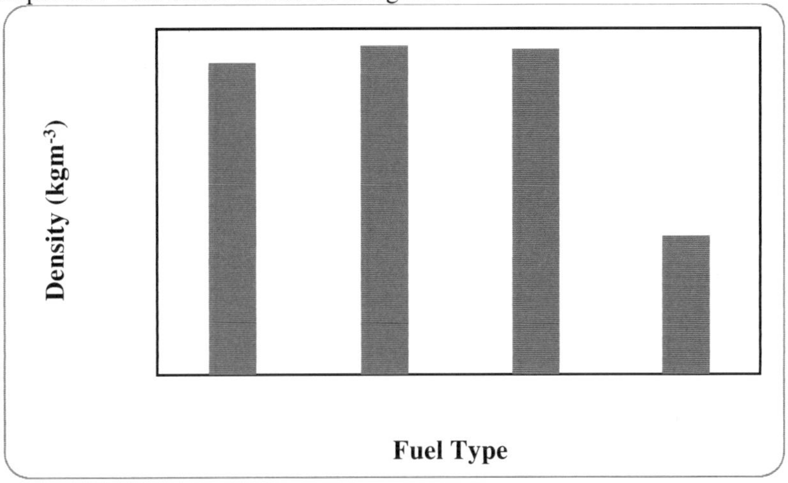

Figure 7. Influence of density for different biodiesel proportions

3.4. *Procedure for Operation of Bomb Calorimeter*
Bomb calorimeter is a machine used to determine the heating value of a given fuel sample. Heating value is defined as the energy content present in a fuel, which is measured by the quantity of heat liberated during the combustion process for a given mass of fuel sample. MPW fuel of 1 gm is weighed accurately and is placed in to crucible. A cotton thread along with fuse wire of certain length is taken by noting down their masses. Bomb head is surrounded with a twisted fuse wire passing through the holes of eyelet. Now one end of cotton thread is tied to the fuse wire and the other end is dipped in the MPW fuel. The crucible along with fuel is placed into a bomb. This bomb is placed in to a container on stands containing 1600ml of water. A thermometer & a stirrer are inserted on top of the container. Thermometer measures the temperature while stirrer maintains uniform temperature. Fill the bomb assembly with oxygen gas and note down the initial temperature of water before testing. Press the fire button to initiate the combustion process of MPW fuel. Note down the final temperature of water to calorific value of MPW fuel. Figure 8.shows bomb calorimeter setup. Repeat the entire procedure to calculate calorific value for PWM and WMP fuels. The longer the chain of the esters higher is the heating value of vegetable oils. Out of these three blends WMP biodiesel has promising heating value but 11.01% lesser than diesel, as heating value depends on carbon chain of esters as shown in figure 9.

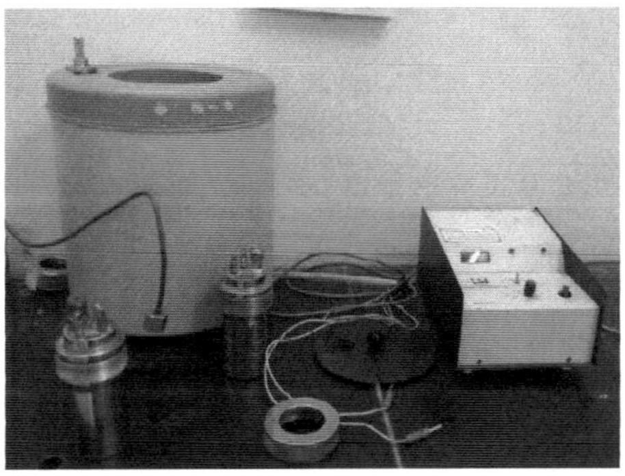

Figure 8. Bomb Calorimeter

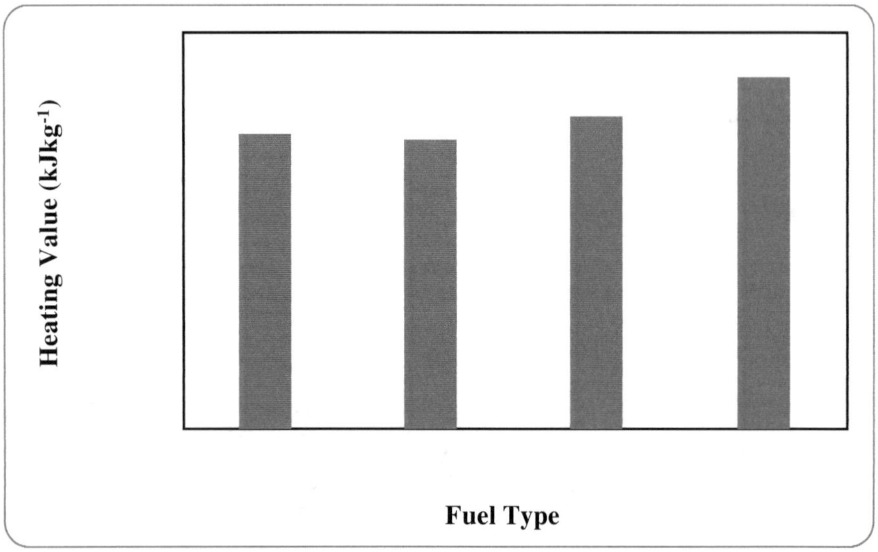

Figure 9. Influence of heating value for different biodiesel proportions

3.5. Estimation of Acid Value

Acid value is an important property of oil as it indicates the free fatty acids present in the sample. The acid value in a fuel is associated with the engine deposits and corrosion of fuel injection parts, so it is important to analyse the acid value of biodiesel. Acid value or neutralisation value for an oil/ fat is the number of milligrams of potassium hydroxide consumed to neutralise per gram of sample. The method followed to determine is simple one, as basically it involves an acid – base titration. The fatty acid groups present in the sample are neutralised by the base that is KOH using phenolphthalein as an indicator. One gram of sample in 30 ml of alcohol is taken and heated for 2 minutes and titrated against alcoholic potash solution as shown in figure 10. The maximum accepted value for biodiesel is 0.5 milligram of potassium hydroxide per gram sample. The acid values obtained for three mixtures are as shown in table 2.

Figure 10. Volumetric analysis

Table 2. Acid values for different fuels

Sl.No	Fuel Type	Acid value(mg KOH/g sample)
1	MPW	0.237
2	PWM	0.280
3	WMP	0.173
4	Diesel	0.6

4. Conclusions

In the present work we tested to explore different possible proportions of biodiesel blends with respect to the fuel properties such as flash point, fire point, density, kinematic viscosity, heating value, acid value of MPW, PWM & WMP biodiesel blends. From the investigation the following conclusions are reported

i. Flash point and fire point temperature of WMP biodiesel is more than that of diesel which does not pose any storage and transportation risks due to due to low volatility.
ii. The MPW biodiesel has better kinematic viscosity compared to other biodiesel blends aiding in proper atomization and good combustion.
iii. Density of MPW, PWM and WMP biodiesels are almost the same. They match with the density of diesel fuel.
iv. Heating value of WMP biodiesel is compatible with that of diesel fuel indicating high energy content that results in the higher brake thermal efficiency.
v. Acid value of biodiesels should be as less as possible. WMP biodiesel acid value is lesser when compared to that of MPW and PWM biodiesels.
vi. Therefore WMP blend can be considered as promising biodiesel which can be further tested for engine performance and emissions.

References

[1] Dario A, Emir S, Pascal C and Romano S D Flash Point, Kinematic Viscosity and Refractive Index: Variations and Correlations of Biodiesel–Diesel Blends *Journal of the Brazilian Society of Mechanical Sciences and Engineering* 2020, **42** 342-47.

[2] John C B, Antony R S, Ranjitha J and Deepanraj B Estimation of Fuel Properties and Characterization of Hemp Biodiesel using Spectrometric Techniques. *Energy Sources, Part A: Recovery, Utilization and Environmental Effects* 2020, 1-17.

[3] Arun S B, Suresh R and Yatish K V Biodiesel Production and Investigation of Fuel Properties. *International Conference on Advanced Materials, Energy & Environmental Sustainability* 2019, **17** 186-90.

[4] Ravichandra D, Ravi K P and Chandramohan V P Production and Compariosn of Fuel Properties for Various Biodiesels. *Lecture Notes in Mechanical Engineering Springer, Singapore* 2018, 267-76.

[5] Acharya N, Nanda P, Panda S and Acharya S Analysis of Properties and Estimation of Optimum Blending Ratio of Blended Mahua Biodiesel *Engineering Science and Technology, an International Journal* 2017, **20** 511-7.

[6] Sania S, Kaisanb M U, Kullab D M, Obib A I, Jibrinc A and Ashok B Determination of Physico chemical Properties of Biodiesel from Citrullus Lanatus Seeds Oil and Diesel /Blends *Industrial Crops & Products* 2018, **122** 702-8.

[7] Mamdouh T G, Mohamed Y E, Aldajah S, Hosam E S and Adel M M Hussien Effect of

Blending on Physiochemical Properties of Jojoba Diesel Fuels *Biodiesels* 2016, **7** 1-14.
[8] Gerald K and Makame M Effects of Biodiesel Blending with Fossil Fuel on Flow Properties of Biodiesel Produced from Non-Edible Oils *International Journal of Green Energy* 2010, **7** 434-44.
[9] Aydin F, Kafadar A B, Erdogan S, Saydut A, Kaya C and Hamamci C The Basic Properties of Transesterified Corn Oil and Biodiesel-Diesel Blend *Energy Sources* 2011, **33** 745-51.
[10] Osmano S V, Vanya M D P, Carlos R P B and Jose R S Physical-Chemical Properties of Waste Cooking Oil Biodiesel and Castor Oil Biodiesel Blends *Fuel* 2010, **99** 1700-2.
[11] Roshan R and Naveen K Performance and Emission Characteristics of Twin Cylinder Diesel Engine Fueled with Mahua Biodiesel and Di-Ethyl Ether *Transportation Engineering* 2020, **2.**
[12] Harisha P, Anil Kumar B N, Tilak S R, Ganesh C Production and Optimization of Biodiesel from composite Ponagmia oil, Animal Fat oil and Waste cooking oil using RSM *Materials Today: Proceedings* 2021, **47** 4901-5
[13] Gaurav D and Sharma M P Investigation and Improvement in Cold Flow Properties of Pongamia Biodiesel *Waste Biomass Valor* 2015, **6** 73-9.
[14] Arulprakasajothi M, Yuvarajan D, Santhanakrishnan R, Suresh V and Beemkumar N Emissions Analysis on Mahua Oil Biodiesel and Higher Alcohol Blends in Diesel Engine, *Alexandria Engineering Journal* 2018, **57** 2627–31.
[15] Abed K A, Morsi A K El, Sayed M M, Shaib A A El and Gad M S, Effect of Waste Cooking-Oil Biodiesel on Performance and Exhaust Emissions of a Diesel Engine, *Egyptian Journal of Petroleum* 2018, **27** 985-9.

Response surface methodology to predict the effects of graphene nanoplatelets addition to diesel engine performance

Sarbani Daud[1,*], Mohd Adnin Hamidi[1, 2], Rizalman Mamat[1]

[1]Faculty of Mechanical and Automotive Engineering Technology, Universiti Malaysia Pahang, Pekan, Malaysia.
[2]Automotive Engineering Centre, Universiti Malaysia Pahang, Pekan, Malaysia

*Email: sarbani@ump.edu.my

Abstract. Engine emissions have become one of the major problems of the world today. Therefore, researchers need to find ways to reduce engine emissions. There are many available methods to reduce emissions or improve engine performance such as using an alternative engine or using alternative fuels. The simplest method is by introducing additives to the currently used fuel and engines. Nanoparticles of zinc oxide, titanium oxide, aluminium oxide, and cerium oxide are among the popular additives used by researchers. The results from the research have been very positive, as it successfully reduced engine emissions. However, there are concerns about the toxicity of the emissions that exposed hazards to human health. A few researchers introduced carbon-based nanomaterials as an additive to improve engine performance and reduce engine emissions. The use of carbon-based nanomaterials is very promising since it poses little to no effect on human health. Graphene is a carbon-based nanomaterial used as an additive in this study. This study aimed to forecast the effects of graphene nanoplatelets' addition to engine performance. The study used torque, power, brake-specific fuel consumption, and brake thermal efficiency as performance parameters. The prediction models consider speed, load, and blend concentration in the calculation. A full-quadratic model with help of Minitab software is used to develop the prediction model. The model is then presented in surface plot and contour plot. The results show that the prediction model agrees with the experimental data with ±10% accuracy. In conclusion, the RSM model of graphene nanoplatelets' effects on diesel engine performance is producible using full quadratic calculation.

1. Introduction

Greenhouse gas emissions have become one of the main sources of global warming, which is a great challenge to the world today. The problem requires governments around the world to put more restrictions on emission regulation which encourages the academic circle to look for sustainable solutions to energy supplies [1].

There are many available methods to reduce emissions or improve engine performance [2]. The methods include engine modifications, combustion enhancement, and the use of treatment components in the exhaust system are examples of these. [3].

A few recent studies focusing on nanoparticles as biodiesel fuel additives show that these additives have a substantial impact on combustion parameters. This effect is attributed to the nanoparticles' high surface area to volume ratio and thermal conductivity. [4, 5]. Nanoparticles of zinc oxide, titanium oxide, aluminium oxide, and cerium oxide are among the popular additives used by researchers in their studies.

In recent years, many efforts taken by researchers to study the behaviour of the various metal oxide application as an additive in diesel fuel and gasoline fuel. Metal-based nanoparticles such as zinc oxide, titanium dioxide, aluminium oxide, and cerium oxide have been introduced into diesel and biodiesel fuel to improve the combustion behaviour of diesel engines and subsequently improve engine performance and reduce engine emissions [6, 7]. Despite the previously mentioned advantages, metal-based nanoparticle additives and metal compounds emitted by combustion by-products are toxic and environmentally hazardous [8-10]. Some studies, for example, have concluded that metal oxide nanoparticles can cause a variety of health problems, including breathing difficulties, lung problems, and skin allergies [11]. Due to the properties of high energy density, high thermal conductivity, and low toxicity, graphene nanoplatelets (GNPs) have the potential to be used as an environmentally friendly fuel additive. [12]. Furthermore, the beneficial properties of GNPs and related carbon nanomaterials have been demonstrated in a variety of other applications such as batteries, chemical sensors, heat transfer, and transparent conductors. [13, 14].

A few researchers have performed experiments to understand the effects of adding GNP to diesel engine performance. The results were very positive as considerable improvements were found in BSFC and BTE. However, a detailed understanding is required. To the best of the author's knowledge, there is still no research to predict the outcome of the performance after the addition of GNP. Therefore, this study was conducted to predict the effects of GNP as a fuel additive to diesel engine performance using the response surface methodology.

2. Materials and Methods

2.1. Fuel Preparation

The GNP is mixed with diesel by a mechanical stirrer at 800 rpm for 15 minutes and directly afterward an Ultrasonication of 24 kHz for 30 minutes (See Fig. 1). The same method was applied by other researchers [4, 15]. According to literature, sonication can break apart particles, and hence, avoid agglomeration (Soudagar et al., 2019). In this test, the Hielscher UP400S was used.

Aside from pure diesel, four samples were blended with graphene. The quantity of the graphene was 25 ppm (D-GNP25), 50 ppm (D-GNP50), 75 ppm (D-GNP75), and 100 ppm (D-GNP100). The amount of graphene in the blend was also used in other literature [16, 17].

2.2. Engine Setup and Test Cycle

The engine used in this experiment is a single-cylinder compression-ignition (CI) engine Yanmar TF120. It is a 4-stroke engine with a compression ratio of 17.7. The power output is rated at 12 hp and the maximum speed is 2400 rpm. In the beginning, the trial was run using pure diesel for a baseline measurement. The engine was run using diesel fuel at five different speeds (900 rpm, 1200 rpm, 1500 rpm, 1800 rpm, and 2100 rpm). By using a dynamometer, six different loads were applied (0%, 20%, 40%, 60%, 80%, and 100%). The run for each setting was repeated three times. The data from computers were downloaded for analysis. In analysing the performance, four indicators were used. The indicators used were brake torque, brake power, brake specific fuel consumption (BSFC), and brake thermal efficiency (BTE). The same experiment was then repeated with all the D-GNP blends.

2.3. Prediction Method

The response surface methodology (RSM) is one of the most commonly used experimental designs for optimization. Because it allows you to assess the effects of multiple factors and their interactions on one or more response variables. This method is widely used by other researchers [18-20] in doing predictions. The response surface methodology (RSM) was performed using the MINITAB software. A step-by-step process is needed to perform the prediction. Firstly, a design of experiment (DOE) had to be set up. A full factorial design was used to create the DOE. There are two types of factorial designs: full factorial and fractional factorial. In each complete trial or replication of the experiments, the full factorial design generates experimental points by utilising all possible combinations of the levels of the factors. Therefore, with a full-factorial design more precise predictions are expected. The three factors input were speed (rpm), load (%), and blend (ppm). Speed had five levels (900, 1200, 1500, 1800, and 2100) rpm, while load had six levels (0, 20, 40, 60, 80, 100) %. Five levels of the Blend (ppm) were divided by five concentrations (0, 25, 50, 75, and 100) ppm. From this input, MINITAB generated a table to fill in.

After running the experiments, the results were filled into the table generated by the software. A full quadratic term was applied. The terms included were speed, load, blend, speed*speed, load*load, blend*blend, speed*load, speed*blend, and load*blend. With this input, the software generated an equation that can be used to predict the performance output based on the factors set up.

Furthermore, the response surface was chosen and the condition for the surface plot was defined. The response is the parameter to predict which are our performance parameters (brake torque, brake power, BSFC, and BTE). Blend (ppm), speed (rpm), and load (%) were put in as factors in the X-Axis and Y-Axis. To present the RSM, a certain speed was chosen. In this study, the surface plots were displayed for the speed of 2100 rpm. The speed of 2100 rpm is chosen because it is considered the maximum speed of the engine able to run safely for a long period. Next, by choosing a surface plot, the same method was applied step-by-step.

3. Prediction of Effects on Performance

3.1. Brake Torque

After preparing a full factorial DOE table, the data acquired from the experiments were input into the table. The software was then required to predict the torque in response to the three different factors (speed, load, and GNP blend) in the experiments. The regression equation suggested for the torque is as follows:

$$Torque\ (Nm) = 1.40 + 0.00537\,S + 0.1629\,L - 0.0077\,B - 0.000002\,S \times S + 0.002014\,L \times L + 0.000014\,B \times B - 0.000038\,S \times L + 0.000003\,S \times B - 0.000072\,L \times B \quad (1)$$

Where, S is the speed in rpm, L is the load in % and B is the GNP blends in ppm

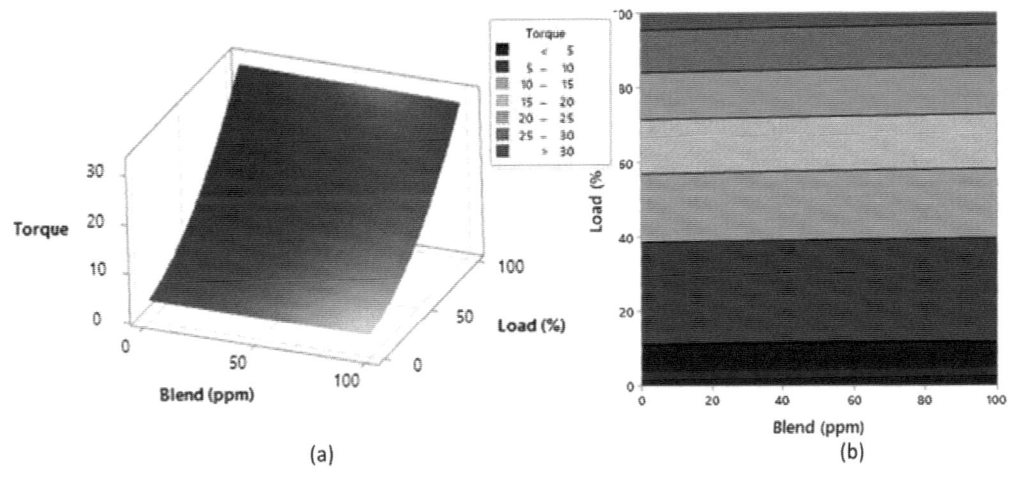

Figure 1. Torque prediction in the form of a) surface plot b) contour plot at 2100 rpm

To simulate the relationship between the torque and the consideration factors, a surface plot and contour plot at 2100 rpm, was created and shown in Figure 1. With 0 ppm GNP, at 0% load, the torque is less than 5 N.m. At 100% load, the torque is more than 30 N.m. With the addition of 100 ppm GNP at 0% load, the torque is less than 5 Nm. At 100 % load, the torque produced is also more than 30 Nm. It can be seen clearly that the difference in load has a big influence on the change of torque value. However, the GNP concentration offers just a marginal effect on the torque. This is in line with research by El-Seesy [21].

3.2. Brake Power

After preparing a full factorial DOE table, the data acquired from the experiments were input into the table. The software was then required to predict the power in response to the three different factors (speed, load, and GNP blend) in the experiments. The regression equation (2) suggested for power is as follows:

$$Power\ (W) = -649 + 1.811\ S - 21.46\ L + 0.35\ B - 0.000557\ S \times S + 0.3230\ L \times L - 0.0013\ B \times B + 0.02457\ S \times L - 0.00028\ S \times B - 0.0122\ L \times B \quad (2)$$

For the surface plot of power vs blend, the load was plotted at 2100 rpm. To get a better reading of the model, a contour plot of those five speeds was also plotted. The results are shown in Figure 2.

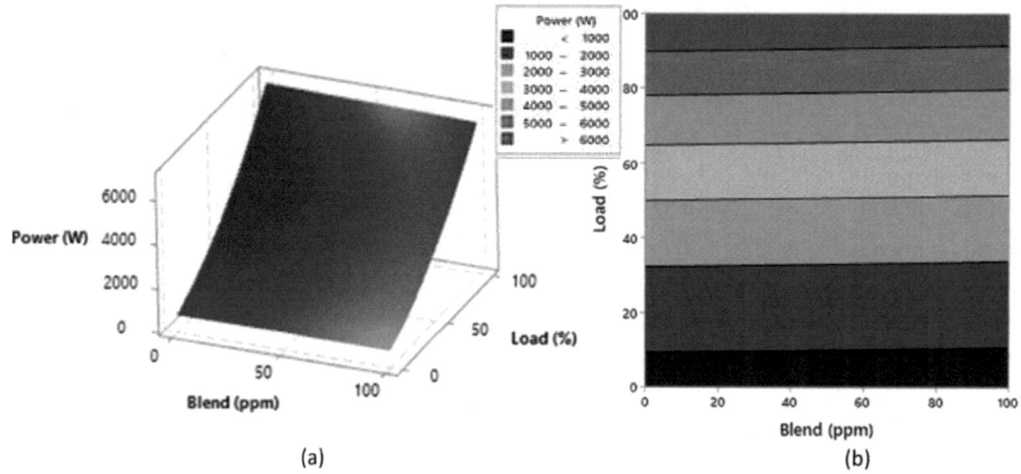

Figure 2. Power prediction in the form of a) surface plot b) contour plot at 2100 rpm

At speed 2100 rpm as shown in Figure 2, with the blend of 0 ppm and load 0%, the power produced is estimated at less than 1000 W. At load 100%, the power to be produced is estimated to be more than 6000 W. Similarly, with D-GNP100 at load 0%, the power to be produced is less than 1000 W and at load 100%, the power will be more than 6000 W. It is clear the parameter load has more influence on the power output. Meanwhile, the power difference between GNP blends is not much to be seen.

3.3. Brake Specific Fuel Consumption

After preparing a full factorial DOE table, the data acquired from the experiments are input into the table. The software was then required to predict the BSFC in response to the three different factors (speed, load, and GNP blend) in the experiments. The regression equation suggested for BSFC is as follows:

$$BSFC\ (g/kW.h) = 394.1 - 0.1352\ S - 1.351\ L + 1.511\ B + 0.000071\ S \times S + 0.01392\ L \times L - 0.00078\ B \times B - 0.000326\ S \times L - 0.000459\ S \times B - 0.01081\ L \times B \quad (3)$$

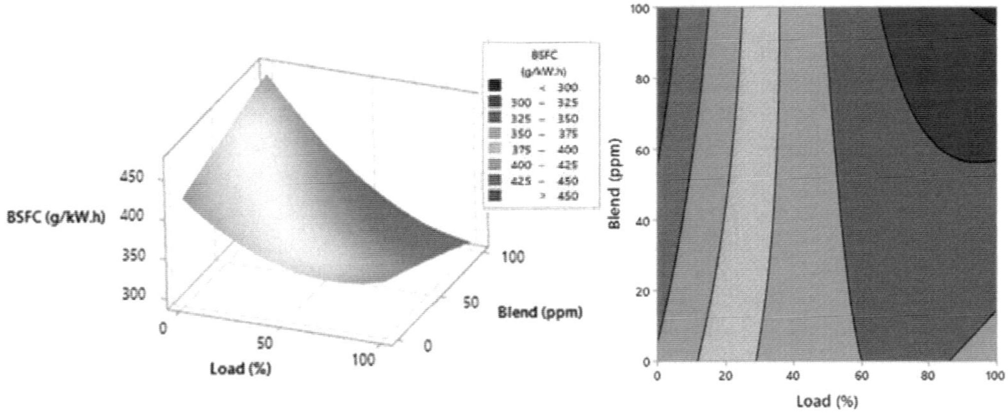

(a) (b)

Figure 3. BSFC prediction in the form of a) surface plot b) contour plot at 2100 rpm

In general, the BSFC is lower at a higher load compared to a lower load. At speed 2100 rpm as shown in Figure 3, 0 ppm GNP, and 0% Load, the BSCF is around 400 – 425 g/kW.h. When the load is raised to 60 – 85 %, the BSFC is about 325 – 350 g/kW.h. When the load is increased to 100%, the BSFC has increased again to 350 – 375 g/kW.h. The reason for this finding is that at a lower load, not all the energy from the blends is used, but the weight of the GNP is included in the BSFC calculation. The addition of GNP increased the difference in BSFC at different loads. The contour is steeper with 8 different steps at 100 ppm GNP as compared to 4 steps at 0 ppm GNP. With 100 ppm GNP at 0% load, the BSFC is more than 450 g/kW.h. With 100 ppm GNP at 100% load, the BSFC is less than 300 g/kW.h. The reason for this is that the energy from GNP is also consumed in this run, therefore GNP not only contributed to the fuel weight but also contributed to the power generated.

3.4. Brake Thermal Efficiency

The result of BTE from the experiment was put into Minitab with the result from the experiment. The regression model equation as suggested by the Minitab software is:

$$BTE\ (\%) = 16.77 + 0.01421\ S + 0.0999\ L - 0.0451\ B - 0.000006\ S \times S - 0.000770\ L \times L - 0.000208\ B \times B + 0.000010\ S \times L + 0.000026\ S \times B + 0.000616\ L \times B \qquad (4)$$

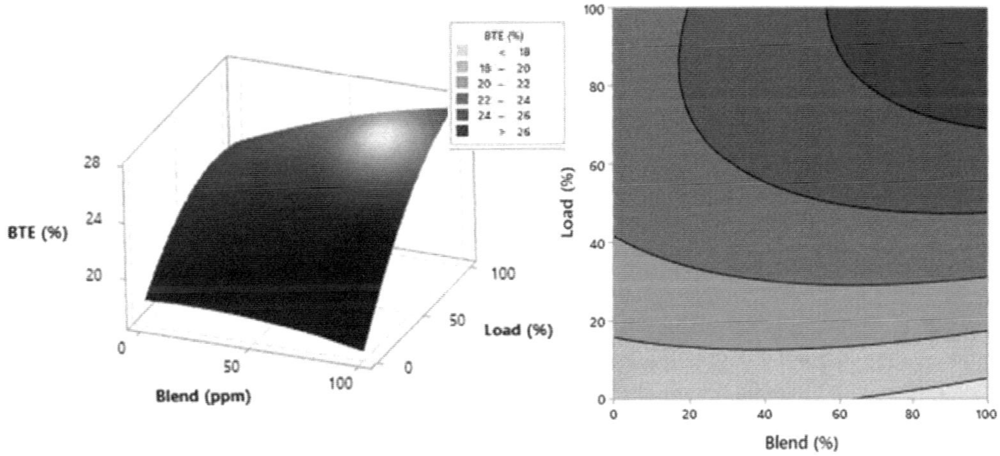

(a) (b)

Figure 4. BTE prediction in the form of a) surface plot b) contour plot at 2100 rpm

At speed 2100 rpm as shown in Figure 4, the difference in loads has more influence on the BTE. It is observed, there is a big difference in the BTE when load 0% is compared to the BTE at load 100%. This is true for all blends. However, note that when GNP is added to the blend, the range of the BTE gets bigger. For example, at load 0% and 0 ppm GNP, the BTE is around 18–20% and at 100% load, the BTE is around 22% - 24%. When 100 ppm of GNP is added, at 0% load, the BTE is less than 18% and at 100% load, the BTE is more than 26%.

3.5. Comparison Between Performance Experimental Value and Model Value

To compare the experimental data against the modelled value, the value of variables which are speed, load, and blend concentration was put into the equations of torque, power, BSFC, and BTE. Using the experimental values and the calculated values, bar charts were created. The value of experimental data and model data are represented side by side, as shown in Figure 5.

The blue bars represent the experimental values while the pink bars represent the model values. From the observation, the experimental values and model values agree with each other where the difference is just less than ±10% only. This acceptance level is also shared by other researchers [22, 23]. This is true for all the performance indicators selected (brake torque, brake power, BSFC, and BTE).

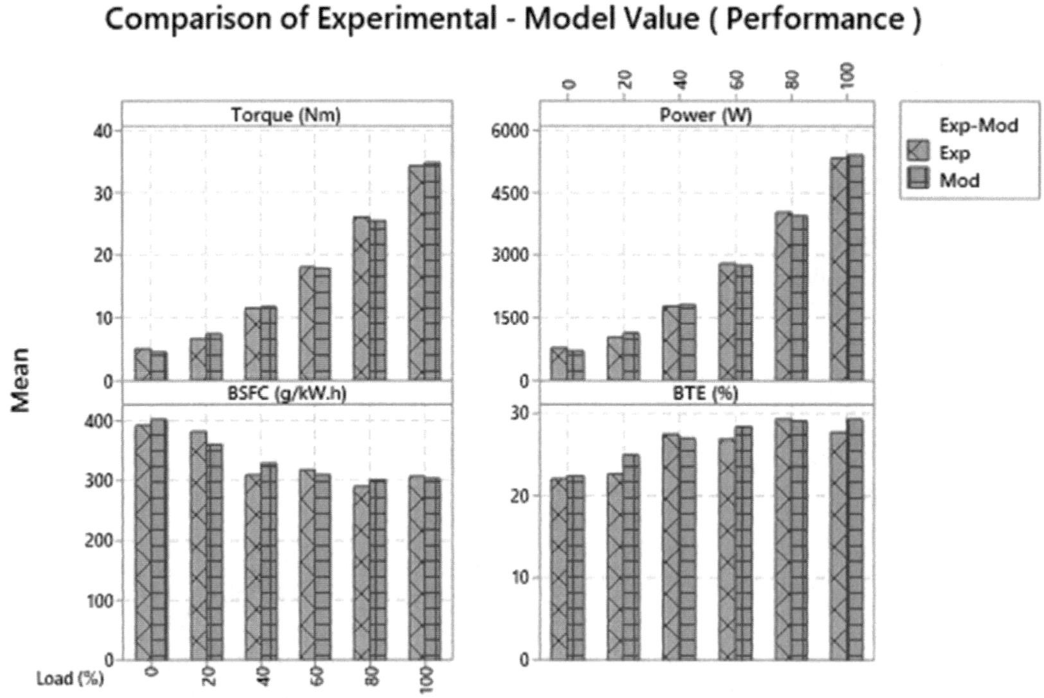

Figure 5. BTE prediction in the form of a) surface plot b) contour plot at 2100 rpm

4. Conclusion

In this study, the effects of the D-GNP blends usage as an additive in a compression-ignition engine have been tried. Predictions models have been developed using the result and statistical software. Model prediction in the form of response surface and contour plot is presented. The prediction models were then tested by comparing the value it suggested with actual experiments' results. The model for all performance parameters (torque, power, BSFC, and BTE) show a good agreement compared to the result from the actual experiment (less than ±10%). Other researchers also share this acceptance level [22, 23].

Acknowledgements
The financial support and laboratory use as offered by UMP Grant, under RDU190358 through Automotive Engineering Centre, Universiti Malaysia Pahang is gratefully acknowledged.

References

[1] M. A. B. Hamidi, M. F. B. Yunadi, and A. Tezaki, "PRF and Toluene/n-heptane Mixture Comparison in HCCI Mode Ignition Using Transient Species Measurements and Simplified Model Analysis, Supported by 0-D and 3-D Simulations," SAE Technical Paper, 0148-7191, 2015.

[2] M. A. B. Hamidi, H. Kosaki, S. Hinata, and A. Tezaki, "Comparison of PRF and toluene/n-heptane mixture in the mechanism of compression ignition using transient species measurements and simplified model analysis," in *8th International Conference on Modeling and Diagnostics for Advanced Engine Systems, COMODIA 2012*, 2012.

[3] M. Aziz *et al.*, "Study of alcohol fuel of butanol and ethanol effect on the compression ignition (CI) engine performance, combustion and emission characteristic," in *IOP Conference Series: Materials Science and Engineering*, 2017, vol. **257**, no. 1: IOP Publishing, p. 012079.

[4] S. S. Hoseini, G. Najafi, B. Ghobadian, M. T. Ebadi, R. Mamat, and T. Yusaf, "Performance and emission characteristics of a CI engine using graphene oxide (GO) nano-particles additives in biodiesel-diesel blends," *Renewable Energy,* vol. **145**, pp. 458-465, 2020/01/01/ 2020, doi: https://doi.org/10.1016/j.renene.2019.06.006.

[5] J. S. Basha *et al.*, "An emission control strategy in a low capacity single cylinder compression ignition engine powered with DEE blended fuels," *Materials Science for Energy Technologies,* vol. **3**, pp. 770-779, 2020/01/01/ 2020, doi: https://doi.org/10.1016/j.mset.2020.09.004.

[6] V. Saxena, N. Kumar, and V. K. Saxena, "A comprehensive review on combustion and stability aspects of metal nanoparticles and its additive effect on diesel and biodiesel fuelled C.I. engine," *Renewable and Sustainable Energy Reviews,* vol. **70**, pp. 563-588, 2017/04/01/ 2017, doi: https://doi.org/10.1016/j.rser.2016.11.067.

[7] K. Nanthagopal, B. Ashok, A. Tamilarasu, A. Johny, and A. Mohan, "Influence on the effect of zinc oxide and titanium dioxide nanoparticles as an additive with Calophyllum inophyllum methyl ester in a CI engine," *Energy Conversion and Management,* vol. **146**, pp. 8-19, 2017/08/15/ 2017, doi: https://doi.org/10.1016/j.enconman.2017.05.021.

[8] M. Horie and K. Fujita, "Chapter Four - Toxicity of Metal Oxides Nanoparticles," in *Advances in Molecular Toxicology*, vol. **5**, J. C. Fishbein Ed.: Elsevier, 2011, pp. 145-178.

[9] D. van der Merwe and J. A. Pickrell, "Chapter 24 - Toxicity of nanomaterials," in *Veterinary Toxicology (Second Edition)*, R. C. Gupta Ed. Boston: Academic Press, 2012, pp. 383-390.

[10] Y. Zhu *et al.*, "Graphene and graphene oxide: synthesis, properties, and applications," *Advanced materials,* vol. **22**, no. 35, pp. 3906-3924, 2010.

[11] I. Izzudin, A. Yusop, S. Sapee, M. Hamidi, I. Yusri, and R. Mamat, "Experimental studies of single cylinder engine run on diesel-biodiesel-butanol blends," in *IOP Conference Series: Materials Science and Engineering*, 2020, vol. **863**, no. 1: IOP Publishing, p. 012060.

[12] J. B. Ooi, H. M. Ismail, V. Swamy, X. Wang, A. K. Swain, and J. R. Rajanren, "Graphite Oxide Nanoparticle as a Diesel Fuel Additive for Cleaner Emissions and Lower Fuel Consumption," *Energy & Fuels,* vol. **30**, no. 2, pp. 1341-1353, 2016/02/18 2016, doi: 10.1021/acs.energyfuels.5b02162.

[13] M. R. Esfahani, E. M. Languri, and M. R. Nunna, "Effect of particle size and viscosity on thermal conductivity enhancement of graphene oxide nanofluid," *International Communications in Heat and Mass Transfer,* vol. **76**, pp. 308-315, 2016/08/01/ 2016, doi: https://doi.org/10.1016/j.icheatmasstransfer.2016.06.006.

[14] Y. Song *et al.*, "Enhancing the thermal, electrical, and mechanical properties of silicone rubber by addition of graphene nanoplatelets," *Materials & Design,* vol. **88**, pp. 950-957, 2015/12/25/ 2015, doi: https://doi.org/10.1016/j.matdes.2015.09.064.

[15] N. Chacko and T. Jeyaseelan, "Comparative evaluation of graphene oxide and graphene nanoplatelets as fuel additives on the combustion and emission characteristics of a diesel engine fuelled with diesel and biodiesel blend," *Fuel Processing Technology,* vol. **204**, p. 106406, 2020/07/01/ 2020, doi: https://doi.org/10.1016/j.fuproc.2020.106406.

[16] S. Debbarma, R. D. Misra, and B. Das, "Performance of graphene-added palm biodiesel in a diesel engine," *Clean Technologies and Environmental Policy,* vol. **22**, no. 2, pp. 523-534, 2020/03/01 2020, doi: 10.1007/s10098-019-01800-2.

[17] N. Singh and R. Bharj, "Effect of CNT-emulsified fuel on performance emission and combustion characteristics of four stroke diesel engine," *International Journal of Current Engineering and Technology,* vol. 5, no. 1, pp. 477-485, 2015.

[18] S. M. Safieddin Ardebili, A. Taghipoor, H. Solmaz, and M. Mostafaei, "The effect of nano-biochar on the performance and emissions of a diesel engine fueled with fusel oil-diesel fuel," *Fuel,* vol. **268**, p. 117356, 2020/05/15/ 2020, doi: https://doi.org/10.1016/j.fuel.2020.117356.

[19] M. Ghanbari, G. Najafi, B. Ghobadian, T. Yusaf, A. P. Carlucci, and M. Kiani Deh Kiani, "Performance and emission characteristics of a CI engine using nano particles additives in biodiesel-diesel blends and modeling with GP approach," *Fuel,* vol. **202**, pp. 699-716, 2017/08/15/ 2017, doi: https://doi.org/10.1016/j.fuel.2017.04.117.

[20] S. Manigandan, P. Gunasekar, J. Devipriya, and S. Nithya, "Emission and injection characteristics of corn biodiesel blends in diesel engine," *Fuel,* vol. **235**, pp. 723-735, 2019/01/01/ 2019, doi: https://doi.org/10.1016/j.fuel.2018.08.071.

[21] A. I. El-Seesy and H. Hassan, "Investigation of the effect of adding graphene oxide, graphene nanoplatelet, and multiwalled carbon nanotube additives with n-butanol-Jatropha methyl ester on a diesel engine performance," *Renewable Energy,* vol. **132**, pp. 558-574, 2019/03/01/ 2019, doi: https://doi.org/10.1016/j.renene.2018.08.026.

[22] S. Simsek, S. Uslu, and H. Simsek, "Proportional impact prediction model of animal waste fat-derived biodiesel by ANN and RSM technique for diesel engine," *Energy,* vol. **239**, p. 122389, 2022/01/15/ 2022, doi: https://doi.org/10.1016/j.energy.2021.122389.

[23] S. Gupta, P. Patel, and P. Mondal, "Biofuels production from pine needles via pyrolysis: Process parameters modeling and optimization through combined RSM and ANN based approach," *Fuel,* vol. **310**, p. 122230, 2022/02/15/ 2022, doi: https://doi.org/10.1016/j.fuel.2021.122230.

IOP Conf. Series: Earth and Environmental Science 1042 (2022) 012004 doi:10.1088/1755-1315/1042/1/012004

Influence of Simarouba Glauca (Laxmi taru) biodiesel on engine performance and emissions of conventional compression ignition engine

Prashant Puthani[1*], Ramesh Jeeragal[1], Amaresha T Patil[1], Malati Chanagond[2]

[1]Department of Mechanical Engineering, BLDEA's V. P. Dr. P. G. Halakatti College of Engineering and Technology, Vijayapur-586103, Karnataka, India

[2]Department of Chemistry, BLDEA's S B Arts and K C P Science College, Vijayapur-586103, Karnataka, India

E-mail: pputhani@gmail.com

Abstract. The present paper reports the performance and emission studies for a four stroke single cylinder CI-DI unmodified engine using laxmi taru biodiesel as fuel. Laxmi taru biodiesel has 80% potential of extracting oil from the seeds as compared to other biofuels. The properties of laxmi taru biodiesel fuel nearly match the properties of diesel fuel. In this paper performance and emission tests are conducted for different blends such as S0 (0% laxmi taru + 100% diesel), S5, S10, S20, S40 by volume. At full load conditions (4.9 kW) brake thermal efficiency (BTE) decreased for S5, S10, S20 and S40 blends by 3.26%, 4.53%, 3.16% and 5.54% respectively as compared to diesel fuel operation because of poor heating value of laxmi taru biodiesel. Presence of oxygen content in the fuel and better combustion leads to decreased carbon monoxide (CO) emission for S5, S10, S20 and S40 blends by 4%, 8%, 12% and 28% as that of diesel. Similarly hydrocarbon (HC) emission reduced by 16.21%, 24.32% and 29.7% in S10, S20 and S40 blends respectively. Oxides of Nitrogen (NO_x) emission increased by 27.19%, 30.57%, 15.2% and 19.3% for S5, S10, S20 and S40 blends respectively in comparison with diesel fuel, under full load conditions due to high incylinder temperature. Usage of laxmi taru biodiesel in CI engines resulting in drastic reduction in carbon based (CO, HC) emissions without altering the performance of the engine. Therefore laxmi taru biodiesel is better alternate fuel for CI engines.

1. Introduction

In recent year's rapid rate of industrialisation, transportation has led to the depletion of fossil fuels along with increase in global warming levels. Even though the emission standards have been revised with new engine modifications the quality of air pollution is not improved [1]. On this earth huge numbers of plant species are available from which the oil extracted from there seeds can be mixed with diesel fuel for use in internal combustion engines. Scientists are experimenting on different plant oils as they produce less emission gases when compared to the diesel fuel, also there is no much difference in the performance of the diesel engines. In the nature there are a number of non-edible oil seeds whose oil can be mixed with diesel to run the vehicles to reduce the global warming effects also unemployment problem will be reduced. Biodiesel extracted from these oilseeds reduces the dependency on foreign fuels to improve the financial economy of a country. Biodiesel increases the life of engine as it has self-lubricating property.

Patil et al. [1] investigated the brake specific fuel consumption (BSFC), BTE, HC, CO, CO_2 and NO_x for various blends of karanja biodiesel along with diesel fuel. Brake thermal efficiency (BTE) is high for full load conditions whereas brake specific fuel consumption (BSFC) is low for part load conditions with less CO and HC emissions. Sathyamurthy et al. [2] conducted test on CI engine using corn oil methyl ester biodiesel blends. They concluded that maximum engine efficiency about 33.98% for B10 blend, BSFC increased for B20 and B30 blend by 4% and 6% respectively in comparison with diesel fuel. Telgane et al. [3] analysed emission and performance parameters of a compression ignition engine fuelled with biodiesel blends of milk scum. At 2.7 kW load for B60 blend BTE lowered by 18.2% with increase of 18.2% in brake specific energy consumption (BSEC) as compared with diesel fuel. Emissions CO decreased by 50%, un-burnt HC reduced by 6.8% with rise in NOx by 33.4%. Verma et al. [4] reported that there is an improvement in BTE when simarouba biodiesel blended with base diesel fuel. Gad et al. [5] reported that waste cooking biodiesel when blended with kerosene and gasoline showed improved results in performance parameters and reduction in emissions.

Gautam et al. [6] conducted test using tallow biodiesel in CI engine and reported that specific fuel consumption is larger in biodiesel blends when compared with diesel fuel due to lower calorific value. Mubarak ct al. [7] reported that there was a maximum reduction in peak pressure with increase in BTE and less BSFC by the use of salvinia-molesta biodiesel in comparison with diesel fuel. Also there was a drop in the emissions when compared with diesel fuel. Ogunkunle et al. [8] investigated that exhaust emissions were lesser than diesel except NO_x for primary biodiesel. Amanpreet Singh et al. [9] conducted test on Jojoba biodiesel blended with ethanol and diesel to estimate the performance parameters as well as engine emissions. There was an increase in BTE with reduction in BSFC and CO emissions. Saravanan et al. [10] carried test on dual biodiesel of rapeseed and mahua in equal proportions by volume. The results showed that at 20% blend, BTE was lesser with decrease in hydrocarbon, carbon monoxide emissions and smoke in comparison with diesel fuel. Bedar et al. [11] conducted experiment on dual cylinder common rail direct injection for Jatropha biodiesel with EGR method to lower NO_x emissions without affecting other emissions. Gorkal et al. [12] conducted experiment on modified low heat rejection engine to increase BTE with decrease in engine emissions. Datta et.al [13] investigated emission and performance parameters of a constant pressure heat addition engine using blend of Jatropha curcas biodiesel and mineral diesel. Improvement in HC and CO emissions was seen with the increase in quantity of biodiesel to diesel. There was rise in NO_x emission by 24% with biodiesel when compared with diesel fuel. Mishra et.al [14] conducted performance and emission characteristics of a compression ignition engine using blend of simarouba glauca up to B20 (20% Simarouba + 80% Diesel). It is concluded that simarouba glauca can be utilized in CI engine as alternative fuel without modification.

From the research papers reviewed it has been observed that the considerable amount of work is carried out on biodiesel using different tree borne oil seeds. Experimental work on utilization of laxmi taru biodiesel in CI engine is scanty. In the past literature, only 20% biodiesel blends were tested for engine performance and emission parameters. Therefore in this paper performance parameters and exhaust emissions are measured and analysed for different blends of laxmi taru biodiesel up to 40%. Finally it will help to estimate the optimal blend without altering the performance of the engine with less harmful pollution to the atmosphere.

Laxmi taru scientifically known as Simarouba Glauca is mainly found in Florida, South America and Caribbean. In India it was introduced in the year 1966 by National Bureau of Plant Genetic Resources (NBPGR), Maharashtra. Now it is cultivated in Orissa, Maharashtra, Gujarat, Rajasthan, Karnataka, Tamil Nadu and West Bengal. Warm, humid, tropical regions are the favourable conditions for laxmi taru cultivation. Temperature range required for cultivation is 10°C-40°C. This evergreen tree height is about 12-15meters with yellow coloured flowers and oval purple coloured fruits. The leaves and bark of laxmi taru tree are used in the treatment of malaria, dysentery, fever and many other medicinal purposes. The seeds have 75-80% of oil, which is extracted by conventional methods. Its oil cakes can be used as manure for plants [15].

2. Materials and Methodology

2.1. Materials

Laxmi taru seeds are been collected from Karnataka State Akkamahadevi Women's University Vijayapur. Oil is extracted from the seeds using oil expeller and then converted into biofuel using methanol and sodium hydroxide or potassium hydroxide by transesterification process. The biofuel is then water washed and supplied for engine test. 1kg of Laxmi taru seeds produce 700-800ml of laxmi taru biofuel which is more in quantity when compared to other oil seeds. The colour of the laxmi taru biodiesel is green. The physical properties of laxmi taru biodiesel match nearly to that of diesel. Table 1. represents the properties of diesel and laxmi taru biodiesel. Similar properties are estimated in the literature [15].

Table 1. Fuel Properties of diesel and laxmi taru biodiesel

Properties	Diesel	Laxmi taru biodiesel
Mass Density [kg/m^3]	820	866
Kinematic viscosity [cSt]	2.41	4.68
Fire point [°C]	68	180
Flash point [°C]	59	168
Calorific value [kJ/kg]	44316	39352

2.2. Experimental Details

The experimental tests are carried out on a 4 stroke, single cylinder, compression ignition engine. A computerised engine test setup layout is shown in figure 1. The specifications of the engine are listed in table 2.

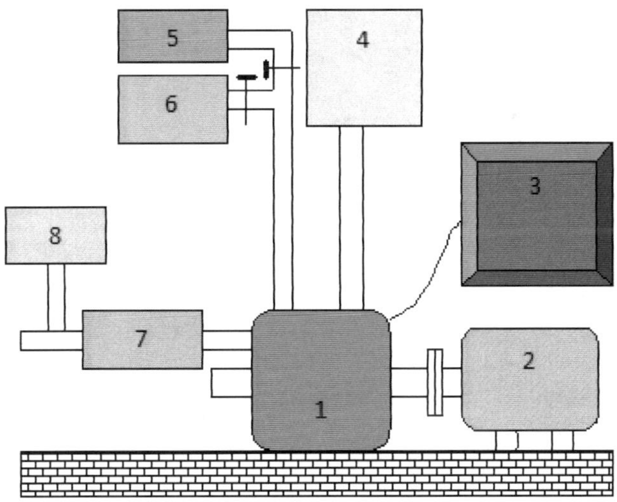

1. Kirloskar make TV1 engine type 2. Eddy current dynamometer with controller
3. Computer for data loging 4. Air box with filter 5. Biodiesel supply tank
6.Diesel supply tank 7. Calorimeter 8. Smoke meter and Gas Analyzer

Figure 1. The layout of the computerised test setup

IOP Conf. Series: Earth and Environmental Science 1042 (2022) 012004 doi:10.1088/1755-1315/1042/1/012004

Table.2: Detailed Engine Specifications

Parameter	**:**	**Specifications**
Engine make	:	Kirloskar
Engine Power	:	5.2kW
Speed	:	1500rpm
Compression ratio	:	17.5:1
Number of cylinder	:	1
Number of strokes	:	4
Bore	:	87.5mm
Stroke length	:	110mm
Injection pressure	:	200 bar
Injection timing	:	23°bTDC
Dynamometer	:	Eddy current type
Cooling	:	Water cooled

2.3. Methodology

Tests are conducted for different blends (S0, S5, S10, S20 and S40) on a compression ignition direct injection engine. Experimental setup of single cylinder CI engine test setup is shown in Figure 2. Engine is operated with neat diesel and laxmi taru biodiesel blends for analysis of performance and emission parameters. Experiments are repeated thrice to obtain accurate data.

Figure 2. Experimental setup of single cylinder CI engine

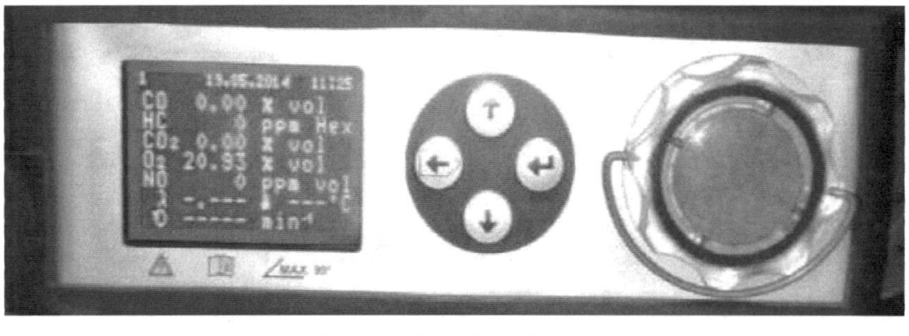

Figure 3. AVL 444 Five Gas Analyser

Fuel consumption was measured on volume basis. Volume flow rate was measured using burette and stopwatch. Time required for the flow of 10 cubic centimetre of fuel was recorded at various loads. Air consumption was measured using U-tube manometer. Eddy current dynamometer was used to put different loads on the engine. All the tests were carried at fixed speed of 1500 rpm. For each loading sufficient time was given for engine stabilisation. During the test fuel consumption, air consumption and temperature of exhaust gas are recorded. Performance characteristics such as BP, BTE, BSFC and exhaust gas temperature (EGT) are recorded and analysed. Also emission parameters such as CO_2, CO, HC and NO_x for different blends are recorded using AVL 444 five gas analyser as shown in figure 3.

3. Results and Discussions

Experimental tests are conducted on constant speed CI engine with laxmi taru biodiesel blends and diesel fuel at various loads. Performance characteristics namely BTE, BSFC, EGT are analysed. Emission characteristics namely CO, CO_2, HC, NO_x are also analysed. The results of performance and emission characteristics are reported below.

3.1 Performance Characteristics

3.1.1 Brake Specific Fuel Consumption (BSFC): Influence of laxmi taru biodiesel (S0, S5, S10, S20 and S40) on BSFC with BP blends is given in figure 4. BSFC is function of fuel consumption rate and the brake power of the engine. It depends on the quantity of fuel injected, density, viscosity and calorific value of the fuel. From figure 4 it can be shown that at low loads BSFC increases with increase in blend percentage of laxmi taru biodiesel in diesel fuel. It is due to high density of laxmi taru biodiesel compared to that of diesel. As load increases BSFC decreases for all blends, it is due to more mass injection of laxmi taru biodiesel which has higher density and lower calorific value. BSFC increased from 0.30 kg/kWhr for diesel to 0.327 kg/kWhr for S40 at 4.9 kW. BSFC is higher about 1.45%, 3.21%, 6.49% and 7.92% for the blends S5, S10, S20 and S40 respectively, when compared with diesel fuel at full load.

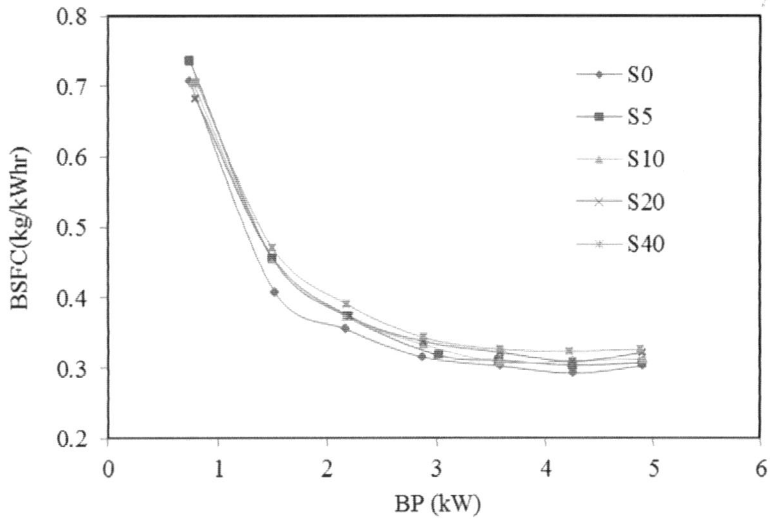

Figure 4. Influence of Laxmi Taru biodiesel on BSFC at various BP

*3.1.2 Brake Thermal Efficiency (BTE):*The alteration of BTE with BP for diesel and laxmi taru biodiesel blends is reported in figure 5. BTE is function of BP and input fuel energy. It is observed that BTE increases with increase in load for diesel and biodiesel blends tested. With increase in the blend proportion BTE decreases throughout when compared to neat diesel. It is because of less energy content available in the biodiesel, high viscosity and poor volatility. BTE decreased from 26.79 % with diesel to 25% with S40 at full load. Therefore, BTE for S5, S10, S20 and S40 blends decreased about 3.26 %, 4.53 %, 3.16 % and 5.54 % respectively when compared with diesel.

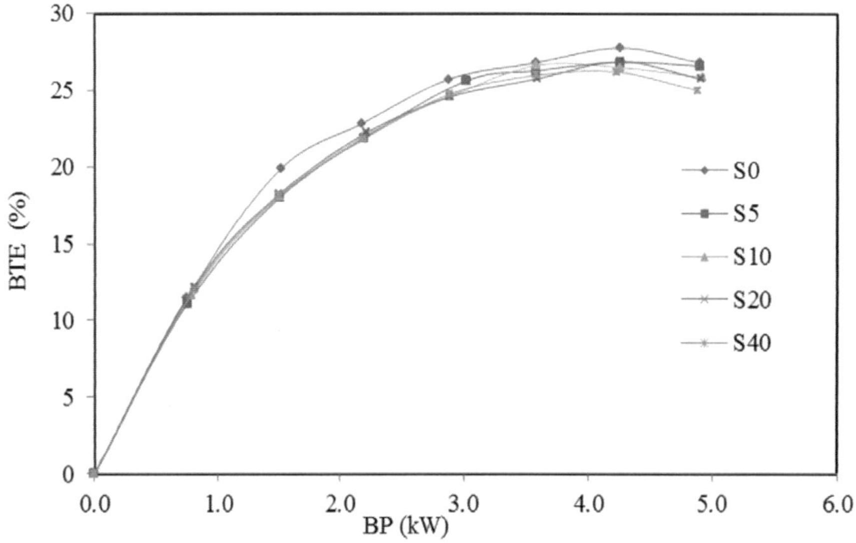

Figure 5. Influence of Laxmi Taru biodiesel on BTE at various BP

3.1.3 Exhaust Gas Temperature (EGT): The change in EGT with BP for neat diesel and laxmi taru biodiesel blends is reported in figure 6. It is seen that EGT rises with addition in load on the engine. At part load condition, there is a slight rise in the EGT of laxmi taru blends when compared to diesel which is due to slow combustion as well as presence of excess oxygen in the fuel. At 4.9 kW S10 and S20 blends EGT are 1.11 % and 1.08 % respectively lower than that of diesel. This is due to good combustion of fuel taking place with reduction in energy loss in the exhaust gas.

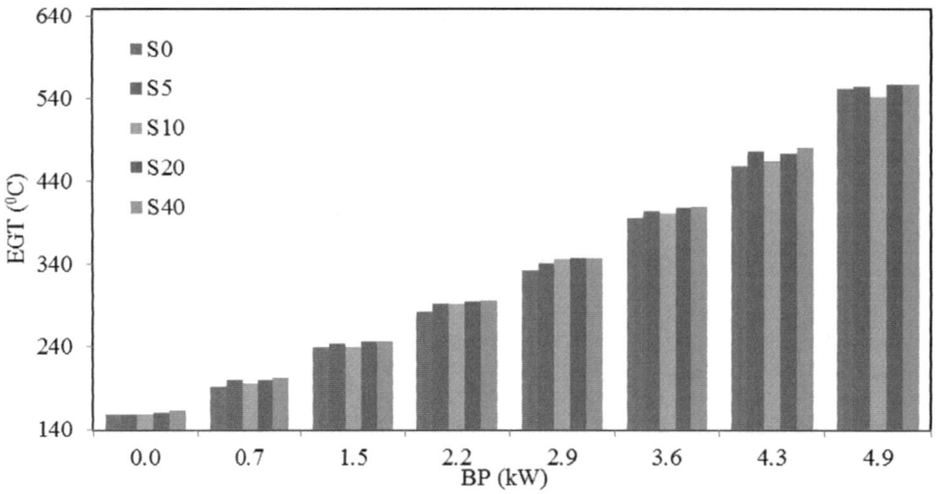

Figure 6. Influence of Laxmi Taru biodiesel on EGT at various BP

3.2 Exhaust Gas Emission Parameters

The emission analysis for neat diesel and laxmi taru biodiesel blends are conducted using exhaust gas analyser.

3.2.1 Carbon Monoxide (CO) Emission: The change in carbon monoxide (CO) emission with BP for neat diesel and laxmi taru biodiesel blends is as given in figure 7. CO is a colourless, odourless, toxic gas developed due to partial combustion of fuel. It is detected that CO emission for laxmi taru biodiesel blends is low when compared to that of neat diesel. It is because of extra oxygen content present in laxmi taru biodiesel helps CO to form CO_2 resulting in complete oxidation of the fuel. At 4.9 kW CO emission for S5, S10, S20 and S40 blends decreased about 4 %, 8%, 12 % and 28% as that of diesel.

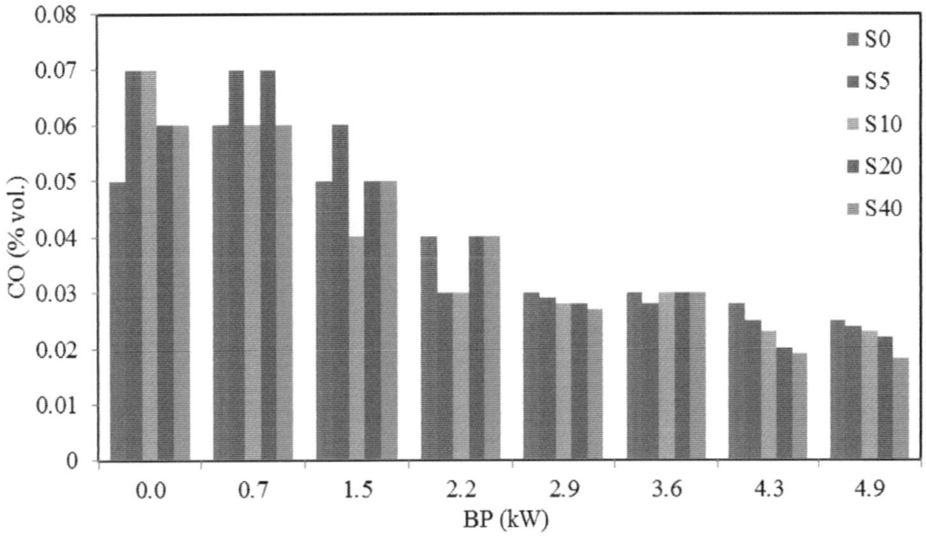

Figure 7. Influence of Laxmi Taru biodiesel on CO at various BP

3.2.2 Carbon Dioxide (CO_2) Emission: The change in CO_2 emission with BP for neat diesel and laxmi taru biodiesel blends is as given in figure 8. CO_2 is a colourless, odourless, nontoxic gas liberated due to combustion of fuel. It is seen that CO_2 emission rises with addition in load as more quantity of fuel enters the combustion chamber. There is a slight rise in CO_2 emission for laxmi taru blends of biodiesel when compared to diesel. The presence of oxygen content in laxmi taru biodiesel improves oxidation of fuel resulting in increased CO_2 emission. At 4.9 kW CO_2 emission for S5, S10, S20 and S40 blends increased about 1.1%, 1.5%, 2.2% and 4% as that of diesel.

3.2.3 Hydrocarbon (HC)Emission: The change in HC emission with BP for neat diesel and laxmi taru biodiesel blends is as given in figure 9. HC is a pollutant generally measured in ppm (parts per million). Hydrocarbon emissions are due to inadequate mixing of air with fuel resulting in incomplete combustion. Hydrocarbon emission rises with addition in load for the entire test. It is noted that laxmi taru biodiesel blends produce less hydrocarbon emission when compared with diesel. This is due to excess oxygen content present in laxmi taru biodiesel blends leads to complete oxidation of the fuel. At 4.9kW there is a reduction of 16.21%, 24.32% and 29.7% of hydrocarbon emission in S10, S20 and S40 blends respectively.

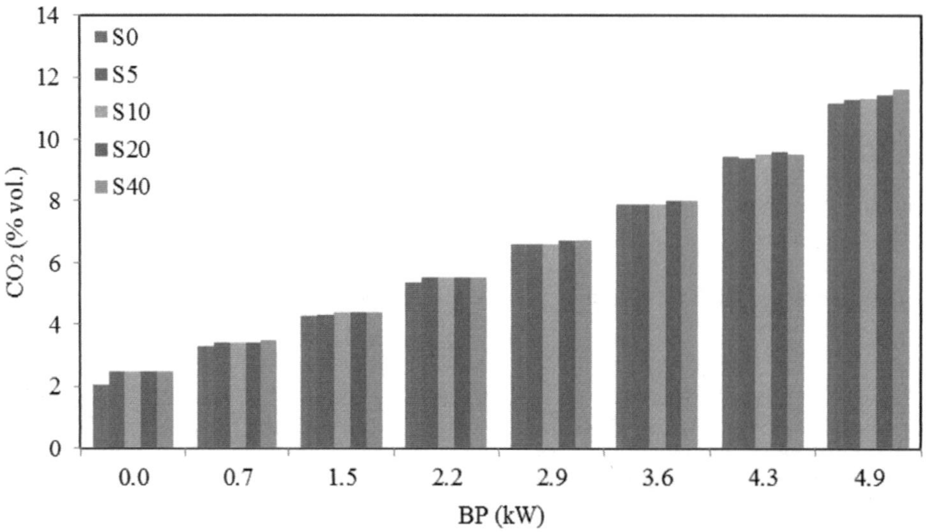

Figure 8. Influence of Laxmi Taru biodiesel on CO_2 at various BP

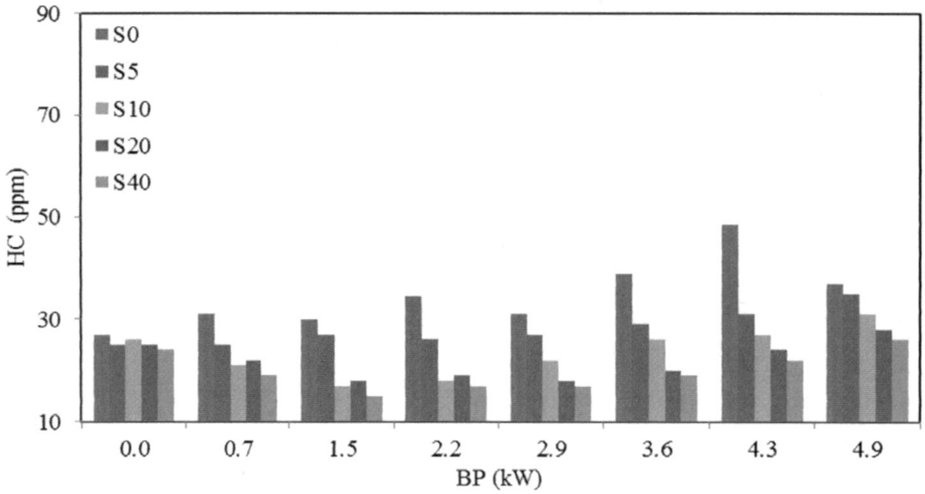

Figure 9. Influence of Laxmi Taru biodiesel on HC at various BP

3.2.4 Oxides of Nitrogen (NO_x) Emission: The change in NO_x emission for neat diesel and laxmi taru biodiesel blends is as shown in figure 10. NO_x is a pollutant which is developed due to higher in cylinder temperature. Oxygen content of the fuel is the main reason for NO_x formation. With increase in load NO_x emission increases for diesel and laxmi taru biodiesel blends. NO_x emission for biodiesel blends increases when compared to that of diesel. At 4.9kW there is increase in NO_x by 27.19%, 30.57%, 15.2% and 19.3% for S5, S10, S20 and S40 blends respectively as compared to that of diesel. Similar results were obtained by Mishra et.al. [15]. They have conducted the experiment and reported the NO_x emissions for B10 and B20 only. In the present investigation the NOx emissions are reported up to 40% biodiesel blend.

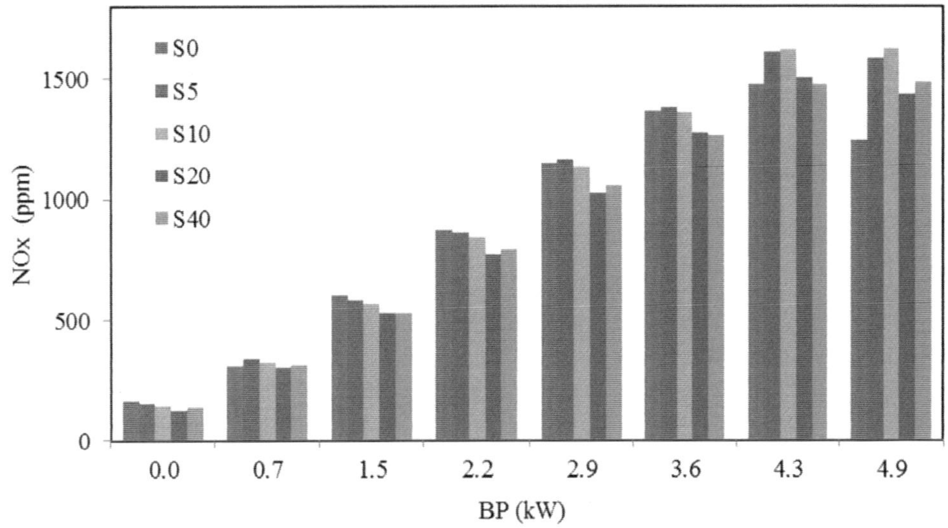

Figure 10. Influence of Laxmi Taru biodiesel on NO_X at various BP

4. Conclusions

Laxmi taru biodiesel is a non-edible, less toxic with more oil content biodiesel. It can be grown in waste lands very easily. Laxmi taru biodiesel has more potential of extracting oil from the seeds as compared to other biodiesel sources. Experiments are conducted on compression ignition engine to investigate the engine performance and emission parameters for neat diesel and laxmi taru biodiesel blends. The following conclusions have been reported.

BSFC increased by 1.45%, 3.21%, 6.49% and 7.92% for S5, S10, S20 and S40 blends respectively when compared with diesel fuel operated at full load conditions. There is also decrease in BTE for S5, S10, S20 and S40 blends by 3.26%, 4.53%, 3.16% and 5.54% respectively when compared with diesel because of lower heating value of laxmi taru biodiesel. Marginal variations in the performance parameters are observed as compared to base diesel fuel operation using various blends of laxmi taru biodiesel. Hydrocarbon and carbon monoxide emissions are reduced with rise in the percentage of laxmi taru biodiesel due to better oxidation. CO emissions at full load conditions for S5, S10, S20 and S40 blends decreased by 4%, 8%, 12% and 28% as that of diesel whereas there is an increase in CO_2 emission for S5, S10, S20 and S40 blends increased about 1.1%, 1.5%, 2.2% and 4% in comparison with diesel. At full load condition, hydrocarbon emission reduces by 16.21%, 24.32% and 29.7% in S10, S20 and S40 blends respectively. NO_x emission increased by 27.19%, 30.57%, 15.2% and 19.3% for S5, S10, S20 and S40 blends respectively when compared with diesel at full load condition.
Laxmi taru biodiesel has 80% potential of extracting oil from the seeds as compared to other biofuels. Usage of S20 biodiesel blend in CI engines resulting in drastic reduction in carbon based (CO, HC) emissions without altering the performance of the engine. Therefore laxmi taru (S20) biodiesel is better alternate fuel for CI engines to minimize conventional energy demand with decreased air pollution.

References

[1] Patil V R, Sane S S and Thipse S S., A Study on Performance and Emission Characteristics of Diesel Engine for Lower Blends of Karanja Biodiesel *Advances in Mechanical Engineering* 2021, **7** 159-67.

[2] Sathyamurthy R, Balaji D, Gorijan S, Jenoris M S, Bharatwaaj R,Vasanthaseelan S and Essa F A, Performance, Combustion and Emission Characteristics of a DI-CI Diesel Engine fueled with

Corn Oil Methyl Ester Biodiesel Blends *Sustainable Energy Technologies and Assements* 2021, **43** 1-10.

[3] Telgane V, Godiganur S, Srikanth H V and Patil S, Performance and Emission Characteristics of a CI Engine Fueled with Milk Scum Biodiesel. *Materials Today Proceedings* 2020, **45** 284-9.

[4] Verma A, Rawat K S, Saify M, Singh A K and Maheshari P, Performance Analysis of CI Engine Powered with Simarouba Glauca L. Biodiesel Fuel *Advances in Materials Engineering and Manufacturing Processes* 2020, 41-50.

[5] Gad M S and Ismail M A, Effect of Waste Cooking Oil Biodiesel Blending with Gasoline and Kerosene on Diesel Engine Performance, Emissions and Combustion Characteristics *Process Safety and Environmental Protection* 2020, **149** 1-30.

[6] Gautam R and Kumar S, Performance and Combustion Analysis of Diesel and Tallow Biodiesel in Engine *Energy Reports*, 2020, **6** 2785-93.

[7] Mubarak M, Shaija A and Suchithra T V, Experimental Evaluation of Salvinia Molesta Oil Biodiesel/Diesel Blends Fuel on Combustion, Performance and Emission Analysis of Diesel Engine *Fuel* 2020, **287**.

[8] Ogunkunle O and Ahmed N A, Exhaust Emissions and Engine Performance Analysis of a Marine Diesel Engine Fuelled With Parinari Polyandra Biodiesel-Diesel Blends *Energy Reports* 2020, **6** 2999-3007.

[9] Singh A, Singh S, Singla V and Singh V, Performance and Emission Analysis of a CI Engine using Ethanol and its Blends with Jojoba Biodiesel and Diesel as Fuel, *Advances in Interdisciplinary Engineering. Lecture Notes in Mechanical Engineering* 2019, 229-40

[10] Sarvanan A, Murugan M, Reddy M S and Parida S, Performacne and Emission Characteristics of Variable Compression Ratio CI Engine Fuelled with Dual Biodiesel Blends of Rapeseed and Mahua *Fuel* 2019, **263**.

[11] Bedar P and Kumar G N, Performance Emission and Combustion Characteristics of CRDI Engine Operating on Jatropha Curcas Blend with EGR *Materials Today Proceedings* 2018,**5** 23384-90.

[12] Gorkal A, Reddy C P and Ravi M U, Experimental Investigations on LHR Single Cylinder Air Cooled Diesel Engine using Laxmi taru Biodiesel and its Blends as Alternate to Diesel Fuel *Journal of the Institution of Engineers* 2017, **99** 513-24.

[13] Datta A, Palit S and Mandal B K, An Experimental Study on the Performance and Emission Characteristics of a CI Engine Fuelled with Jatropha Biodiesel and its Blends with Diesel *Journal of Mechanical Science and Technology* 2014, **28** 1961-66.

[14] Mishra S R, Mohanty M K, Panigrahi N and Pattanaik A K, Impact of Simarouba glauca biodiesel blends as a fuel on the performance and emission analysis in an unmodified DICI engine, *Renewable Energy Focus* **26**, 2018, 11-16.

[15] Patil M S and Gaikwad D K, A Critical Review on medicinally important oil yielding Plant Laxmi Taru (Laxmi taru Glauca DC) *Journal of Pharmaceutical Sciences and Research* 2011, 1195-1213.

IOP Conf. Series: Earth and Environmental Science 1042 (2022) 012005 doi:10.1088/1755-1315/1042/1/012005

Characterization of Chia seed oil methyl ester as an alternate fuel for diesel engine

Manjunath S[1], Ramakrishna N Hedge[2*]

[1] Shri Madhwa Vadiraja Institute of Technology & Management, Bantakal, India
[2] Srinivas Institute of Technology, Mangaluru, India

*Email: rkhegderk@gmail.com

Abstract. The present world is questing for low carbon fuel, which can replace diesel and biodiesels are one such promising fuel. In this study, the feasibility of Chia seed oil as a blended bio-diesel for use as partial replacement to Diesel is evaluated through transesterification process. 10 grams of dried, moisture free Chia seeds were subject to transesterification process using methanol as a replacement alcohol. NaOH was used as a catalyst. The resulting mixture was stirred for one hour at 700 rpm, maintaining 70oC. The methyl ester was separated after one hour settling period and washed several times using distilled water at 100oC and dried at 70 ^{o}C and stored. Impurities were removed by Decalcification process that resulted in 80% yield. Characterization of the oil was used different blends from B05 - B25 in an incremental steps five percent by volume with diesel. Major fluid and thermodynamic properties like viscosity, density, specific gravity, etc. were extracted as per the ASTM prescribed standards. The results show that the flash and fire points of Chia seed oil are respectively 2.96 times and 2.57 times higher compared to Diesel, while the kinematic viscosity and density are comparable but 6.88% and 5.2% higher. Further investigation shows that biodiesel blend B20 could be a second fuel for diesel engine.

1. Introduction

World population is growing at an exponential rate which is increasing the demand for energy. For human welfare and economic growth, energy is the major contributing factor. Petroleum based fuels serve as a major conventional energy source, but it is depleting due to limited availability. Hence, researchers are on the quest for alternate fuels that can replace the fossil fuels. A lot of research work is being conducted from the past two decades to extract biofuel from edible oil, non edible seeds, algae, and waste cooking oil as well as waste industrial oil. Coconuts are grown in large quantity in South India, also biodiesel extracted from coconut oil can also be blended with diesel. Transesterification process is a process which involves alcohol reacting with oil in the presence of catalyst which leads to formation of biodiesel. Out of many issues associated with transesterification process; one such problem is alcohols and oils are not readily reactive due to the chemical structure. To overcome this problem, curcas seed oil was trans-esterified with cosolvent- tetrahydrofuran [1].
Currently, biodiesel cost is expensive, which can be solved by using two solutions, firstly, using low priced non-edible oils, waste oils and second, reducing the cost of processing. Two step process to convert Jatropha curcas L. Oil to biodiesel by ultrasonic transesterification was done by Xian den en, at. Using H2SO4 and NaOH as catalyst, a yield of 92.8% and 47.2% was obtained. Stable biodiesel was obtained by pre-treatment of H2SO4 for one hour and acid decreased to 1.2 mg KOH/g from

10.45[2]. Another investigation of biodiesel production from Karanja oil has been made using cement and mortar material as catalyst. Catalysts were prepared by cleaning and calcined for 3 hours at 850°C and characterised by X-ray diffusion. Economic analysis of production process reveals that biodiesel cost was evaluated to be $1.23 per kg [3].

Study and optimization of process can lead to reduction of production cost. An attempt was made to optimize the biodiesel production process from Karanja, Polanga and Jatropaha oils. Biodiesel produced were blended with diesel and evaluated [4].

An experimental investigation was conducted study use of waste animal bone as catalyst for transesterification process. Dynamic light scattering and thermogravimetric analysis, catalyst was characterized, and hydrothermal reaction was used to increase reaction surface area. 96% and 88% yield were obtained from the hydrothermally treated sample and calcined sample respectively [5]. Waste eggshells are used as heterogeneous catalyst distinguished by powder X-ray diffraction and transmission electron microscopy analysis. 96.74% yield was obtained from the 4-hour reaction, temperature of 65°C was maintained and 5 wt.% catalyst loading [6].

For commercialization, the price of biodiesel should be deceased, Crucial step is to use low-cost raw material. Waste cooking oil (small in percentage) is one of the environmental pollutants, which can convert into useful product. Calophyllum inophyllum oil and waste cooking oil mixture was transesterified by microwave irradiation, Yield of 97.40% was obtained with 7.15 min reaction time and 600 rpm stirring speed with higher oxidation stability (18.03h) [7]. Transesterification of algae is known as biodiesel production of third generation. A lot of problems are associated with this process, requirement of large area for algae growth and the time required is also high. Last decade, A lot of research work is being done to solve the issues, one such attempt was nano catalytic transesterification process - biodiesel derived from algae. Using calcium methoxide Ca (OCH3)2 (3wt.%), yield of 99.0% was obtained at 80°C with reaction time of 3hr [8].

Teo et.al. [9], tried to optimize the production of algae and characterize the biodiesel extracted from Spirulina, Pond Water Algae (PWA) and Chlorella, three different algae species. Process techniques like proximate and elemental analysis, gas chromatographic analysis, NMR and FTIR were utilized to study biomass and algae oil. Maximum growth was observed with Chlorella and moisture content was higher compared to Spirulina algae and PWA. Biodiesel extracted form PWA, and Spirulina algae had same fuel properties as that of conventional biodiesel and it was concluded that the same can be a substitute to petroleum diesel fuel [9].

Quest for the biodiesel source is continued, using homogeneous catalyst and ultrasound, biodiesel extraction from Annona squamosal seed was attempted. Maximum yield of 97.6% was achieved with reaction temperature 57°C and 113 min reaction time. GC-MS, TGA and FT-IR were utilized to characterize the transesterification process and according to ASTM standards, fuel properties of Annona squamosal oil were characterized [10]. Edible oils are not feasible for biodiesel production owing to expensive prices. Waste oils and non-edible oils are best suited for large scale biodiesel production. An investigation to extract oil from tamarind seed and convert to biodiesel by transesterification using catalyst HCL acid was reported by Sundaramahalingam et.al [11] utilizing biofuel from tamarind seed having 65-68 cetane number and 26-32 low iodine value with subsequent analysis has been done using GC-MC [11]. Another method of blending of biodiesel for better results from researchers has been reported in recent years that used double blends. Simarouba and Cotton seed oils were blended in equal quantity and characterization of various properties like density, viscosity, pour point and ash content were completed [12]. Finding a potential biodiesel, which can substitute diesel fuel is still on. In this study, the transesterification of Chia seed oil and characterization of its properties is being done.

2. Biodiesel production

In this section, preparation of the Chia seed oil for transesterification is discussed. Transesterification process was done in either single stage or double stage based on results of the free fatty acid test. First the oil was subjected to the free fatty acid test (FFA). Single stage process is taken up, if content of free fatty acid FFA is less than 4%. A process having two stages is undertaken if FFA of raw oil is more than 4%. It is particularly important that the oil is subjected to the FFA test.

Procedure to determine the FFA of raw oil:

Step 1: NaOH (0.1N) 0.1 normal solution preparation. One liter of water was taken in a conical flask, 4 grams of NaOH were weighed and transferred into the conical flask. Water was constantly stirred to dissolve NaOH completely.

Step 2: Titration and calculation of Free fatty acid content in raw oil. 50 ml of isopropyl alcohol was taken in a clean and dry 250ml conical flask. Few drops of 0.1 NaOH solution were added to the flask and the solution was well shaken for proper mixing.

Step 3: 10 grams of Chia seed oil were then added to the above conical flask. Five drops of Phenolphthalein were added as an indicator, followed by thorough mixing of the content. The solution prepared was titrated using 0.1N-NaOH solution, till turned to pink color. In the next step the free fatty acid content in the oil was found using the relation 1.

$$FFA = \frac{28.2 \times Normality\ of\ NaOH \times Titration\ value}{weight\ of\ the\ oil} \quad (1)$$

$$FFA = \frac{28.2 \times 0.1 \times 15}{10} = 4.23$$

As the ingredients of an oil is not just oleic acid, but more than that, the results obtained from equation 1 has inherent errors of small magnitude, which is acceptable.

As FFA > 4, to ensure it within the range as prescribed by ASTM standards, in the next step 1.25ml of H_2SO_4 is added (from the chart, for the value of FFA- 4.23, 1.25ml of H_2SO_4) to reduce the FFA in esterification process [1, 2].

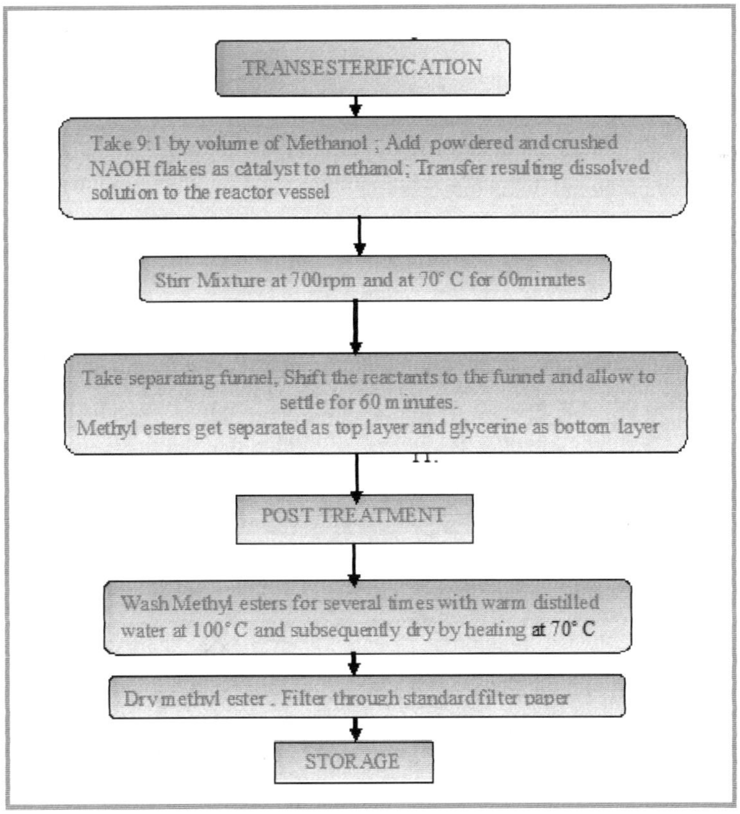

Figure 1: Flow diagram of Methyl ester synthesis

2.1. Transesterification
Alcoholysis is another name for the transesterification process, in which, replacement of alcohol is done from an ester by another alcohol. Butanol, propanol, methanol, and ethanol are suitable alcohol for replacement. The fatty acid methyl ester is obtained, and viscosity of triglycerides decreases, which in-turn improves fuel properties and engine performance. This process yielded 80% of biodiesel. As shown in Figure 1, flow chart explains the synthesis of biodiesel. Figure 2 (a) shows the pictorial representation of transesterification process.

Steps to be followed in transesterification process.
1. Molar ratio of oil to Methanol: 1:9
2. Mass of sodium hydroxide catalyst: 2%(w/v)
3. Temperature of reaction: 70°C
4. Time of reaction: 60 minutes

2.2. Decalcification
After transesterification process, the next step is to remove the impurities. Take a separating funnel and transfer the biodiesel to the same vessel and allowed to settle for 3 hours. Figure 2 (b) shows the impurities settled at the bottom. Impurities were drained from the stopcock. Chia seed oil methyl ester obtained was thoroughly washed with pure water for 10 times [3-5].

3. Characterization of biodiesel
After decalcification, fuel properties are evaluated in the laboratory like kinematic viscosity (KV), specific gravity, density, flash, and fire point. For fuel operational conditions and storage, viscosity is the prime fuel property to be determined.
Figure 2 (c) shows the viscometer with cannon-Fenske tube (tube no 100, direct type). The test bath temperature was kept at 40°C for an interval of 30 minutes and KV was calculated as follows.

$$Kinematic\ viscosity = number\ of\ seconds \times standard\ factor\ of\ the\ bulb\ viscosity \quad (2)$$

$$Kinematic\ viscosity\ of\ Chia\ seed\ oil\ methyl\ ester = 137 \times 0.0238 = 3.26\ cst$$

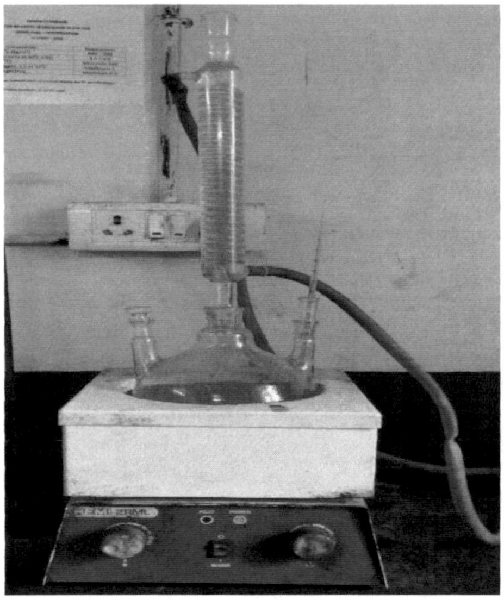

(a) Tran-esterification

(b) Extracted Biofuel from Chia seeds

(c) Kinematic viscometer bath instrument with Cannon-Fensky tube.
Figure 2: Characterization of biodiesel fuel

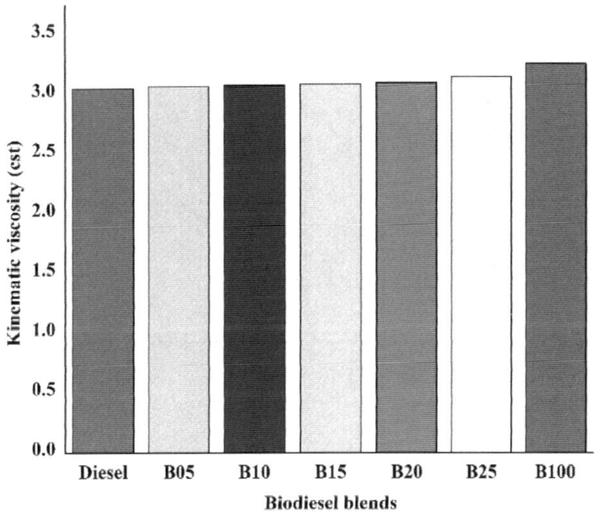

Figure 3: Kinematic viscosity of different bio-diesel blends

Figure 3 depicts the contrast of the KV of different biodiesel blends at 40°C which is evaluated by Cannon-Fenske viscometer (tube no 100, direct type) [7-10]. Table 1 shows the comparison of chia seed methyl ester with pure diesel.

Table 1: Properties of fuels

Properties	Diesel	Chia seed oil methyl ester
Flash point (°C)	56	165
Fire point (°C)	---	167
Kinematic viscosity (cst) at 40°C	3.05	3.26
Density (kg/m3) at 40° C	820	865

IOP Conf. Series: Earth and Environmental Science 1042 (2022) 012005 doi:10.1088/1755-1315/1042/1/012005

Using a glass hydrometer (measurement range 0.7 to 1specific gravity, temperature 4-90°C), a density measurement of the biodiesel and its blends were conducted. The test samples B05, B10, B15, B20 and B25 were prepared by blending Chia seed methyl ester with diesel having volumetric ratios of 5%, 10%, 15%, 20% and 25% respectively. Table 2 shows the specific gravity of biodiesel blend and diesel.

Pensky-Martin flash point instrument is shown in the figure 4. The lowest temperature of momentary ignition of vapor of combustible oil can be evaluated by using this apparatus. Figure 5 shows the contrast of the flash point for different biodiesel blends. The flash point of the Chia seed oil methyl ester is 165°C whereas that of diesel is 57°C. Table 3 depicts the properties of the Chia seed biodiesel, which are within the limits prescribed by the ASTM standards [12, 13].

Figure 4: Pensky Flash point Apparatus

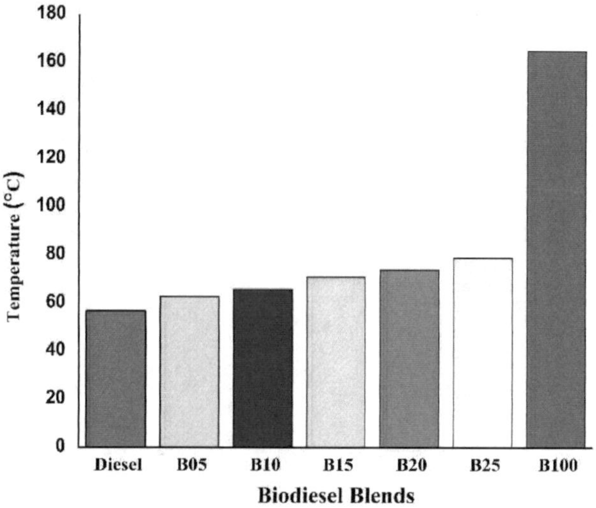

Figure 5: Flash point of biodiesel blends

Table 2. Measured properties of the fuel blend & Diesel

S No	Biodiesel	Specific gravity	Density(kg/m^3)
1	B05	0.822	822
2	B10	0.825	825
3	B15	0.827	827
4	B20	0.829	829
5	B25	0.831	831
6	Diesel	0.820	820

Table 3. Comparison of fuel properties with ASTM standards

Properties	Standard	Range	Experimental value
Flash point (°C)	ASTM D93	>150	165
Viscosity (m^2/s) at 40°C	ASTM D445	$1.9\text{x}10^{-6} - 6\text{x}10^{-6}$	$2.81\text{x}10^{-6}$
Specific gravity	ASTM D4052	0.87-0.90	0.865

4. Conclusions

In this experimental investigation, the properties of the biodiesel resulting from the Chia seed oil were assessed as a second fuel to CI engine. On the basis of the characterization of used Chia oil blends, the following conclusions were made.

- Chia seed oil has flash and fire points respectively 2.96 times and 2.57 times higher compared to Diesel.
- The flow and fluid properties, namely kinematic viscosity and density are comparable, but 6.88% and 5.2% higher than Diesel, making it a potential bio-diesel blend.
- B20 could be a potentially good alternate fuel for a diesel engine on economy angle, with no much sacrifice on the fluid and flow properties.
- From ASTM prescribed standards, properties of the blend B20 are well within the prescribed range, and it justifies the above conclusion.

References

[1] Mohammed-Dabo, I.A., Ahmad, M.S., Hamza, A., Muazu, K. and Aliyu, A., 2012. *Journal of Petroleum Technology and Alternative Fuels*, *3*(4), pp.42-51.
[2] Deng, X., Fang, Z., and Liu, Y.H., 2010. *Energy Conversion and Management*, *51*(12), pp.2802-2807.
[3] Kumar, D., Singh, B., Banerjee, A. and Chatterjee, S., 2018. *Journal of cleaner production*, *183*, pp.26-34.
[4] Sahoo, P.K. and Das, L.M., 2009. *Fuel*, *88*(9), pp.1588-1594.
[5] Chingakham, C., Tiwary, C. and Sajith, V., 2019. *Catalysis Letters*, *149*(4), pp.1100-1110.
[6] Soudagar, M.E.M., Nik-Ghazali, N.N., Kalam, M.A., Badruddin, I.A., Banapurmath, N.R., Ali, M.A.B., Kamangar, S., Cho, H.M. and Akram, N., 2020. *Renewable Energy*, *146*, pp.2291-2307.
[7] Borah, M.J., Das, A., Das, V., Bhuyan, N. and Deka, D., 2019. *Fuel*, *242*, pp.345-354.
[8] Milano, J., Ong, H.C., Masjuki, H.H., Silitonga, A.S., Chen, W.H., Kusumo, F., Dharma, S. and Sebayang, A.H., 2018. *Energy conversion and management*, *158*, pp.400-415.
[9] Teo, S.H., Islam, A., and Taufiq-Yap, Y.H., 2016. *Chemical engineering research and design*, *111*, pp.362-370.
[10] Nautiyal, P., Subramanian, K.A. and Dastidar, M.G., 2014. *Fuel Processing Technology*, *120*, pp.79-88.
[11] Sundaramahalingam, M.A., Karthikumar, S., Kumar, R.S., Samuel, K.J., Shajahan, S., Sivasubramanian, V., Sivashanmugam, P., Varalakshmi, P., Syed, A., Marraiki, N. and Elgorban, A.M., 2021. *Fuel*, *291*, p.120195.
[12] Hanifa, M.A., Mushtaqa, A., Zubairb, M. and Rashidc, U. 2018, *i-SAMN, p16-17*
[13] Hegde, R.N. and Bantwal, J., 2015. *Asian Journal of Engineering and Technology*, *3*(4).

IOP Conf. Series: Earth and Environmental Science 1042 (2022) 012006 doi:10.1088/1755-1315/1042/1/012006

Design and fabrication of a partial flow dilution tunnel for particulate mass sampling in an ethanol blended CI engine

Atul Kumar, Tomesh Kumar Sahu*, Pravesh Chandra Shukla
Department of Mechanical Engineering, Indian Institute of Technology Bhilai, India

*Email: tomeshs@iitbhilai.ac.in

Abstract. Diesel engine is mostly preferred engine especially for heavy duty work, as it delivers high efficiency, durability and low operating cost. Oxides of nitrogen (NOx) and particulate matter (PM) are the main emission components emitted from these diesel engines. Alcohols or biodiesels blending with mineral diesel are known to reduce the NOx and PM emissions. Ethanol fuel contains 35% (m/m) of oxygen, which helps to minimise the smoke and PM emissions. The primary objective of this work is to in-house design and fabricate a partial flow dilution tunnel for PM sampling on substrates. Ethanol blends (E05 and E10; v/v) were tested for Particulate matter collection for a twin cylinder CRDI Compression Ignition Engine. Ethanol blends showed 33% reduction in PM emission compared to mineral diesel.

1. Introduction

Research on alternative fuels has been underway since the 1970s, when a worldwide energy crisis made oil scarce. The report found that biofuels (ethanol and biodiesel) can actually replace petroleum. Diesel engines are preferred for heavy duty activities owing their efficiency, durability and low operational cost. It brings various benefits; however, it is also responsible for environmental pollution [1, 2]. Globally, policies defined in some of the environmental events like the Kyoto protocol aim to limit the environmental impact of diesel fuel related emissions. Diesel engines emit Oxides of nitrogen (NOx) and particulate matter (PM) in significant concentration [3-5]. These PM emissions are detrimental to humans and ecosystems. Delhi had over ten times the WHO target for annual PM2.5 emissions. Therefore, monitoring pollution statistics in India is also quite important. Therefore, India's emission regulation Bharat Stage regulates CI and SI engine pollution. It was initially released in 2000. India is now following emission standard BS VI. Poor air quality is terrible for economy and harmful for Indian citizens. The world is focusing on replacing diesel fuel with bio-fuel, either fully or partially, to reduce pollution and increase energy security. India imports crude oil from Arab countries to meet its energy needs. It is a big economic load on our country. In 2018-19, India spent $111.9 billion to buy 228.6 million tonnes of crude oil [6]. Diesel exhaust gas contains mostly PM and NO_x. More than 50% of pollutant is NO_x. PM is the second most polluting after NO_x. Soot forms at high temperatures and low oxygen levels. Phases of soot creation include particle generation and growth. The capacity of exhaust gas to contain hydrocarbon vapour increases with temperature. The condensate vapour deposited over the soot helps for particle growth. Particulate matter is made up of 31% elemental carbon, 14% sulphates and moisture, 7% unburnt fuel, 40% unburnt lubricating oil, and the remaining portion could be metals or other substances [7]. PM emissions are formed as a result of a number of factors, including the combustion and expansion processes, the quality of the fuel (sulphur and ash content), the quality and consumption of lubrication oil, the combustion temperature, and exhaust gas cooling [7]. Diesel particles are so small that they can penetrate the respiratory tract of an animal or a human and accumulate in the lung's pulmonary area, where they can cause adverse health impacts. The presence of millions of diesel particulate particles per cubic centimetre also enters the bloodstream. Long-term employment in industries that expose workers to diesel exhaust may result in a 20–50% increase in the chance of suffering lung cancer. Apart from that, diesel particulates have been shown to increase the risk of blood cancer also [8]. In this context, over the last decades, the use of numerous alternative fuels has been encouraged [9,10].

Another advantage of alternative fuels is that many of them are environmentally friendly and produce significantly less CO_2. Stricter emission regulations have been implemented, promoting the use of alternative fuels such as alcohols, which are known to emit less particulate matter. Approximately 60%–70% of transport engines are diesel engines, and alcohol combustion in CI engines has gained popularity in recent years. Particulate emission is a significant regulated emission in a CI engine and it is composed of a large number of organic compounds, many of which are toxic. Gangwar et al. [11] conducted experiments on a four-cylinder, water-cooled, inline CI engine running on diesel and a 20% biodiesel blend at various loads and speeds of 1800 and 2400 rpm [11]. Particulate mass increases with engine load for both diesel and diesel blends (B20), and diesel has a slightly higher particle mass than biodiesel blends [11]. Betha et al. [12] conducted experiments in a single-cylinder CI engine that's been fuelled with ULSD, pure biodiesel (used cooking oil), and 50 percent used cooking oil in ULSD fuel. It was determined that as the loading on the diesel generator increased, the particle mass concentration for all of the fuels increased too. Overall emission of particulate matter from each blend was found to be lower than that of ULSD [12]. Agarwal et al. [13] tested a four-cylinder direct injection diesel engine with mineral diesel and a 20% Jatropha biodiesel blend (B20). The engine ran at 1800rpm under five different loads. It produced more particles for diesel compare to B20 fuelled condition. also, they reported that particulate mass rose with load [13]. Zhu et al. [14] conducted an experiment using a direct-injection diesel engine and Euro V diesel fuel containing diethyl adipate. DEA is present in the fuel blend in concentrations of 8.1 percent, 16.4 percent, 25 percent, and 33.8 percent by volume fraction, equivalent to 3 percent, 6 percent, 9 percent, and 12 percent by mass of oxygen in the blend with Euro V diesel. While particle mass concentration increases with engine load, it decreases with increasing DEA in the fuel blend. In comparison to Euro V diesel fuel, the measured particle mass concentrations are lower for tested fuels [14]. Chen et al. [15] tested a 4 cylinder engine using pure diesel and diesel/ethanol blends. The blend was formed by blending 10%, 20%, and 30% of ethanol into pure diesel. They found that low and high engine loads produce more PM in the exhaust and the addition of ethanol to fuels lowered PM emissions [15]. Using a four-stroke multi-cylinder diesel engine, Yusop et al. [16] tested pure diesel and a blend of diesel, ethanol, and biodiesel in a compression ignition engine for particle emission. The amount of ethanol in diesel and biodiesel blends ranges from 5% to 20%. When ethanol content in blended fuels is increased, mass concentration of PM decreases for low and medium loading condition [16].

It was noted that use of biofuel especially alcohols addition into diesel help to control the particulate formation and a sampling instrument is necessary to characterise the PM emission from test fuel. This device can be used to collect particulate matter from exhaust exiting the tail pipe. The current work is to design and fabricate a particle sampling instrument called partial flow dilution tunnel and characterise emissions of twin cylinder compression ignition engines running on mineral diesel and lower ethanol blends of mineral diesel (E5, E10).

2. Design and Methodology of Partial Flow Dilution Tunnel Fabrication

Dilution tunnel is a sampling device for collection of Particulate matter form the exhaust gas which simulate the condition when exhaust comes in environment and dilute with ambient air. A small portion of the exhaust gas is diluted in the partial flow dilution systems and the entire diluted sample is then passed via the filters. This tunnel is equipped with quartz filter paper having pore size 2200 nm, fitted in particulate assembly portion to filter the diluted exhaust gas which pass through dilution tunnel. Design method and philosophy for fabricating each component has been discussed in this section.

2.1 Design Condition and Philosophy
To replicate automobile-like surroundings, exhaust flue gas is mixed with atmospheric gas. The dilution ratio should be somewhere around 5 -25. Equation 1 illustrates the relationship used to determine the dilution ratio. It is assumed that when exhaust gases exit the tailpipe and come into contact with ambient air, it dilutes by a factor of 5 - 25, and the temperature of the diluted gas must be greater than 293K to avoid the thermophoresis effect. The experiment used a flow rate of 3 m^3/min. The dimensions of the heating chamber, mixing chamber, suction pump capacity, air blower capacity, and orifice plate capacity, among others, have been designed accordingly. Additionally, a particulate sampling system has been integrated into the partial flow dilution tunnel to collect particulate matter. The tunnel should be suitably long before to the sample zone to facilitate thorough mixing of the dilution air and exhaust gas. To avoid flow separation, the Reynolds number should be greater than 4000. The background CO_2 concentration of air intake is neglected due to the high concentration of exhaust CO_2 in equation 1.

$$\text{Dillution ratio} = \frac{\text{Undiluted exhaust CO}_2}{\text{Diluted exhaust CO}_2 - \text{Ambient air CO}_2} \quad (1)$$

2.2 Design of Dilution Tunnel
The dilution tunnel is made up of six distinct sections, each of which is developed and manufactured individually as indicated in figure 1. The components are as follows: inlet manifold, air blowing Chamber, mixing chamber, exhaust manifold, particle assembly, and orifice plate. The Stainless Steel ERW Pipes that were employed in the construction of this tunnel for durability.

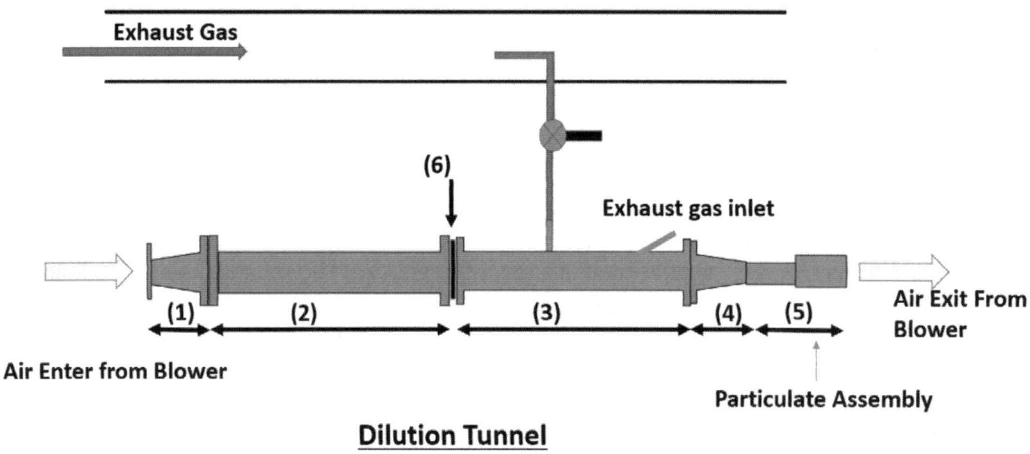

Figure 1. Schematic of the partial flow dilution tunnel and different parts: (1) Inlet manifold (2) air blowing chamber (3) Mixing chamber (4) Exhaust manifold (5) Particulate assembly (6) Orifice plate

2.2.1 Design of Inlet manifold. Initially modelling of Partial Flow Dilution Tunnel is performed using CAD modelling software. While designing the inlet manifold, it should ensure that there should not be any flow separation (Sparrow et al. 2009) [17]. Simulation performed by sparrow et al. mentioned that a diffuser angle of 10 degrees for that flow separation will not occur at Reynold number 2000-33000. Figure 2 shows the actual figure of inlet manifold after fabrication where diffuser angle of 10 degree is adopted and experiment is performed beyond 4000 Reynold number, in order to avoid separation phenomenon.

2.2.2 Design and Fabrication of Air Blowing Chamber. The air blowing chamber shown in figure 3 is used to direct the incoming air from the input manifold into the proper direction. This pipe is connected to the input manifold by a flange, and it is also connected to the orifice plate on another side. It measures 1066.8 millimeters in length and 101.6 millimeters in diameter.

Figure 2. Developed inlet manifold with dimensions

Figure 3. Developed Air blowing chamber

2.2.3 Design of Mixing Chamber. Figure 4 shows the mixing chamber, it is similar in length and proportions to an air blowing chamber. Three probes, were drilled in this chamber. The first drill hole was drilled to accommodate the manometer probe, which is used to detect pressure. A second drill hole was drilled to allow for the introduction of the exhaust gas sample for mixing. A hose pipe was utilised to transport a portion of engine exhaust to the mixing chamber. While the third drill hole was used to collect diluted exhaust samples for CO_2 monitoring in order to calibrate the tunnel's dilution ratio. CO_2 concentration is determined using an emission analyser probe. The experiment is carried out using a dilution ratio of 10.

Figure 4. Actual image of mixing chamber

Figure 5. Orifice plate fitted between two flanges of pipe.

2.2.4 Design and Fabrication of Orifice Plate. Orifice plate is clamped between two flanges of air blowing and mixing chamber. According to BS 1042 Part 1 1964 [18], pressure vent is located at a distance D and D/2 from orifice plate, where D is the diameter of pipe in which orifice plate is holded. The coefficient for an orifice plate is about 0.52 to 0.58, while the head loss is normally 60% to 70% of the differential pressure. Diameter of stainless-steel pipe is 101.6 mm; therefore, inner diameter of orifice is taken D/2 50.8mm. Outer diameter of Orifice plate is 155mm. with a thickness 10mm.

IOP Conf. Series: Earth and Environmental Science 1042 (2022) 012006 doi:10.1088/1755-1315/1042/1/012006

2.2.5 *Design of Particulate Assembly*. It is fitted with Quartz filter paper. It is designed and produced in a leak-proof manner. Particulate assembly is composed of four distinct components: a sleeve, a nipple, and two rings (shown in figure 6). The sleeve section is threaded on the inside, and an inner ring is attached to it to secure the other ring without leakage. Through internal threading, the sleeve is connected to the nipple, and a quartz filter paper is sandwiched between the two rings. To prevent the filter paper from moving, it is maintained by a steel net with the same dimensions as the ring. The exact image of these components is shown below. The particle assembly is connected to the mixing chamber on one end and to the blower suction component on the other.

Figure 6. Sleeve, nipple and rings used for dilution tunnel

3. Particle sampling for E05 E10 and D100 and Result analysis

The designed partial flow dilution tunnel (figure 7) was placed with the Twin Cylinder CRDI CI engine test rig (figure 8) and particulate sampling experiments were conducted for the fuelled conditions D100, E05, and E10. The engine was operated for 1600 rpm at 40 Nm engine loading condition and particle sampling were performed after reaching the steady state. Particulate sample is collected for 30 min duration for each test fuel i.e., Pure Diesel, E05 and E10 at dilution ratio 10 which is maintained throughout experiment. Particulate matter is sampled using quartz microfiber air filter paper. Its pores size is 2.2 micron and diameter are 47 mm. Filter paper is placed in a desiccator for 24 hours to extract all moisture. Prior to doing the experiment, weighing has done for the filter paper in a weighing balance. After that, it is attached to the particle sampling unit. Following the experiment, it is placed back in the desiccator to eliminate any remaining moisture. Weighing has been done once more for filter paper; the difference in weight represents the mass of particulate matter. Figure 9 shows the result obtained from the performed experiments. For pure diesel fuel, particulate collection in filter paper is 3 mg. After blending with ethanol 5% (v/v) in pure diesel particulate collection is 2 mg which is lower than that of pure diesel. This decrease in particulate collection is due to the presence of inherent oxygen in ethanol, which assist the combustion process of carbon in the fuel.

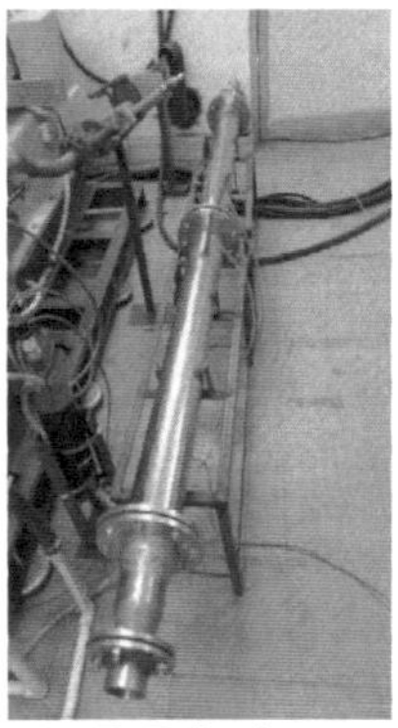

Figure 7. Dilution tunnel

Figure 8. Engine set-up with dilution tunnel

Figure 9. Quartz micro pore filter paper before and after collection of particulate matter

Further blending 10% ethanol (v/v) in pure diesel also tested for particle collection showed in figure 10, it is observed lower than that of pure diesel however it is equal to particulate collected at E05. There is not much difference between inherent oxygen concentration of E05 and E10. Hence, the effect of E05 is significant over the diesel fuel. If further increase the blending to 10% this effect is not significant to further reduce the particulate composition. Beyond 10% increase in blending may reflect some significant decrease as compare to E05.

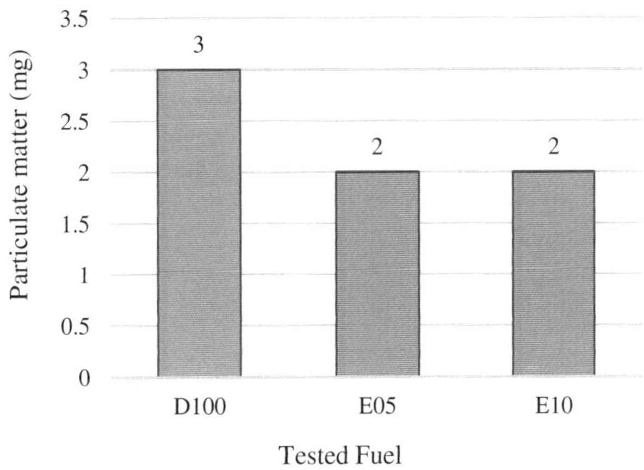

Figure 10. Collected PM (mg) for D100, E05 and E10 at 40Nm engine load @1600 rpm

4. Conclusions

This work mainly involves two aspects. First, the design and in-house fabrication of a partial flow dilution tunnel for test engine setup and second, particulate sampling for diesel and lower blends of ethanol diesel (E05 and E10). Based on the performed experiment following conclusion can be made.

- A partial flow dilution tunnel was designed and fabricated for the purpose of particulate matter (PM) sampling. Experiments were carried out with the tunnel for PM sampling using 47mm quartz filter substrates. It was found that the dilution ratio of the tunnel was able to maintain as required and exhaust and air flow were maintained properly. There was no leakage from the system and PM sampling was performed. Overall, fabricated dilution tunnel demonstrated the sampling of PM.
- Ethanol E05 shows 33% reduction of particulate matter compare to diesel D100 for tested condition. Completed combustion of ethanol and inherent oxygen content and lower carbon to hydrogen ratio help to increase the degree of combustion and control the formation of particulate.

References

[1] Hora T S, Shukla P C and Agarwal A K 2016 Particulate emissions from hydrogen enriched compressed natural gas engine *Fuel* **166** 574-80
[2] Sahu T K, Kshatri R and Shukla P C 2021 Impact of Ethanol on Combustion, Performance, and Emission Characteristics of Diesel Engine - Alcohol as an Alternative Fuel for Internal Combustion Engines ed P C Shukla, G Belgiorno, G D Blasio and A K Agarwal (Singapore: Springer) pp 251-266
[3] Galvão E S, Santos J M, Lima A T, Reis Jr N C, Orlando M T D A and Stuetz R M 2018 Trends in analytical techniques applied to particulate matter characterization: a critical review of fundaments and applications *Chemosphere* **199** 546-68
[4] Shukla P C, Gupta T and Agarwal A K 2014 A comparative morphological study of primary and aged particles emitted from a biodiesel (B20) vis-à;-vis Diesel Fuelled CRDI Engine *Aerosol and Air Quality Research* **14** 934-42
[5] Sahu V K, Sahu T K, Shukla P C 2022 Fuel Injection Strategies for Alcohol Utilization in Combustion Engines - Application of Clean Fuels in Combustion Engines. Springer, Singapore, pp 55-69.
[6] Ersoy E, Schaffer M E and Ditzen J 2019 BP Statistical Review of World Energy 2019
[7] Kittelson D B 1998 Engines and nanoparticles: a review *Journal of aerosol science* **29** 575-88
[8] Raza M, Chen L, Leach F and Ding S 2018 A review of particulate number (PN) emissions from gasoline direct injection (GDI) engines and their control techniques *Energies* **11** 1417
[9] Sahu T K, Gupta S and Shukla P C 2020 Second Generation Bioethanol Production from Organic Waste - Alternative Fuels and Their Utilization Strategies in Internal Combustion Engines ed A P Singh, Y C Sharma, N N Mustafi and A K Agarwal (Singapore: Springer Singapore) pp 49–64
[10] Sahu T K, Sarkar S and Shukla P C 2021 Combustion investigation of waste cooking oil (WCO) with varying compression ratio in a single cylinder CI engine *Fuel* **283** 119262
[11] Gangwar J N, Gupta T and Agarwal A K 2012 Composition and comparative toxicity of particulate matter emitted from a diesel and biodiesel fuelled CRDI engine *Atmospheric environment* **46** 472-81
[12] Betha R and Balasubramanian R 2011 Emissions of particulate-bound elements from stationary diesel engine: Characterization and risk assessment *Atmospheric Environment* **45** 5273-81
[13] Agarwal A K, Shrivastava A and Prasad R K 2016 Evaluation of toxic potential of particulates emitted from Jatropha biodiesel fuelled engine *Renewable Energy* **99** 564-72
[14] Zhu R, Cheung C S, Huang Z and Wang X 2011 Regulated and unregulated emissions from a diesel engine fueled with diesel fuel blended with diethyl adipate *Atmospheric Environment* **45** 2174-81
[15] Chen H, Shi-Jin S and Jian-Xin W 2007 Study on combustion characteristics and PM emission of diesel engines using ester–ethanol–diesel blended fuels *Proceedings of the Combustion institute* **31** 2981-9
[16] Yusop A F, Yasin M H M, Mamat R, Abdullah A A and Aziz A 2013 PM emission of diesel engines using ester-ethanol-diesel blended fuel *Procedia Engineering* **53** 530-5
[17] Sparrow E M, Abraham J P and Minkowycz W J 2009 Flow separation in a diverging conical duct: Effect of Reynolds number and divergence angle *International Journal of Heat and Mass Transfer* **52** 3079-83
[18] Hobbs J M and Humphreys J S 1990 The effect of orifice plate geometry upon discharge coefficient *Flow Measurement and Instrumentation* **1** 133-40

Studies on the catalytic hydrothermal liquefaction of sugarcane bagasse to bio-oil, over cobalt oxide

Gopalakrishnan Govindasamy*, Rohit Sharma and Sunu Subramanian

Department of Chemical Engineering, School of Engineering, University of Petroleum & Energy Studies(UPES), Dehradun-248007, India.

*Email: gopalakrishnan@ddn.upes.ac.in

Abstract. Sugarcane bagasse, a by-product of sugar industry is a potential lignocellulose biomass for bio-oil production by hydrothermal liquefaction (HTL). The catalyst and the reaction parameters are important for the higher bio-oil yield. Hence, by the sol-gel method, cobalt oxide catalyst was synthesized, characterized by FTIR, X-ray diffraction, nitrogen adsorption and studied for the HTL of sugarcane bagasse with varying reaction time, catalyst/biomass weight ratio and water/biomass weight ratio. Formation of face-centered cubic spinel of Co_3O_4 was confirmed from the XRD peaks at 2Θ positions of 18, 31, 37, 45, and 65° and FTIR absorption bands at 558 and 654 cm^{-1} attributed to stretching vibrations of Co-O, having the Co^{3+} in octahedral and tetrahedral coordination respectively. A high BET surface area of 413.63 cm^2/g obtained may be attributed to the citric acid added during sol-gel synthesis, which got decomposed off during the calcination. The highest yield of bio-oil was found to be 57.6% at the reaction time of 120 min, catalyst/biomass ratio of 0.4, water/biomass ratio of 28, 250°C and initial CO pressure of 45 bars, compared to 35.2% for non-catalytic HTL, confirmed its catalytic activity.

1. Introduction

Strong reliance on fossil fuels to meet the global energy demand resulted in its fast depletion, and global warming [1]. Biofuels are an important option to address the twin issues of energy security and mitigation of global warming [2], as they are renewable besides being carbon neutral from the perspective of life cycle analysis [3]. Reviews on the development of biochemical [4] and thermochemical processes [5] for the production of biofuels from lignocellulose, the most abundant non-edible biomass were reported. Globally, 540 million metric tons per annum of sugarcane bagasse, a non-edible lignocellulose biomass is generated as a byproduct in the sugar industries [6]. While it is currently consumed as an in-house fuel for steam generation, its use for the production of second generation ethanol by biochemical process, a substitute for gasoline was reviewed by Bazerra et al [6]. From the green-house gases emission perspective, biochemical process of ethanol production was reportedly better compared to its direct burning [7]. Microbial process for the conversion of sugarcane bagasse to hydrogen involving consortium of microorganisms incubated at different temperatures was reported by Soares et al [8].

Among the thermochemical conversions of sugarcane bagasse, steam gasification and its mechanism were reported by Ahmed et al [9]. Treedet et al reported the fast pyrolysis of sugarcane bagasse to bio-oil in a circulating fluidized bed reactor [10]. Hydrothermal liquefaction (HTL) to produce bio-oil from sugarcane bagasse was studied using homogeneous alkali catalysts viz. KOH, K_2CO_3 which increased the conversion and bio-oil yield by 10-15% compared to non-catalytic HTL

[11]. A study with the heterogeneous catalyst of iron-cobalt oxide, reported an increase in the bio-oil yield by 22% compared to non-catalytic HTL [12]. Formation of bio-oil by the HTL of sugarcane bagasse first involves the depolymerization of lignocellulose to water soluble oxygenates (WSO) such as phenols and sugars by acid-base catalyzed hydrolysis, followed by the deoxygenation of WSO to bio-oil. Hot compressed water used in HTL was reportedly having ionic product 500 times as that of ambient water, thus the H^+ and OH^- furnished by it, acted as acid and base catalysts respectively [13]. The second step and thus the bio-oil yield depends upon the activity of the externally added catalyst and the process parameters [12]. Soni et al reported that the cobalt oxide catalyst was effective for the deoxygenation of jatropha and palm oil to hydrocarbons [14].

Hence, in the present study synthesis, characterization and activity of cobalt oxide as catalyst for the HTL of sugarcane bagasse under different values of process parameters viz. catalyst/biomass weight ratio, reaction time, water/biomass weight ratio at 250°C and initial CO pressure of 45 bar is reported.

2. Materials and Methods

Cobalt nitrate hexahydrate, citric acid monohydrate, isopropanol of analytical grade of Rankem make, diethyl ether, acetone, and ethyl acetate of LR grade of SRL make, carbon monoxide of ultrahigh pure grade were used in the study as such.

2.1. Synthesis and characterization of catalyst

10 g of Cobalt nitrate was dissolved in 40 ml of distilled water and continuously stirred for 2 h at 50°C. To this solution, 2 g of citric acid monohydrate dissolved in 30 ml of isopropanol was mixed. The temperature was raised to 85°C and stirred for 2 hours. Dark precipitate formed was filtered, dried at 140°C for 1.5 hours and calcined for 1 hour at 350°C.

The XRD of the calcined cobalt oxide was recorded using Bruker D8 Advance diffractometer with nickel-filtered Cu Kα radiation (λ = 1.54 A). Textural characteristics were determined by nitrogen adsorption at -196°C using Micromeritics ASAP-2020. Perkin Elmer Frontier model FTIR spectrometer was used to record the spectra with the KBr pellet of the catalyst. SEM-Zeiss EVO-40 EP was used to get the SEM image of the catalyst by giving conductive gold coating on the catalyst sample.

2.2. HTL of sugarcane bagasse

Using a batch reactor with a magnetic stirrer (BR 100 Berghof), sugarcane bagasse was studied for HTL and the schematic diagram of the reactor arrangement used was as already reported [15]. In a non-catalytic reaction run, distilled water of 70 ml was added to the reactor followed by addition 2.5 g of sugarcane bagasse under stirring. The reactor was first purged, then pressurized with CO as the process gas to 45 bar, and sealed. With magnetic stirrer, contents of the reactor were agitated at 1400 rpm. The reactor was heated to 250°C in 12–13 min, at the rate of 20°C/min, while the pressure reached 90 bar. These reaction conditions were kept for 120 min, and then the reactor was brought to the ambient temperature. Gaseous products from the reactor were passed through the CO incinerator and then vented out. Rest of the product was separated and quantified as Oil 1, Oil 2, water soluble organics (WSO), Oil 3, Solid Residue (SR) using the procedure reported earlier [12]. Conversion, yield of the products, and yield of Total Oil were calculated using the equations reported thereof. Amount of catalyst, water and the reaction time were varied in order to study the influence of catalyst/biomass (C/B), water/biomass (W/B) weight ratios and the reaction time respectively.

3. Results and Discussion

3.1. Characterization of catalyst

XRD peaks observed at 2Θ of 18, 31, 37, 45 and 65°, were indexed to (1 1 1), (2 2 0), (3 1 1), (4 0 0), and (4 4 0) planes respectively of the face centred cubic spinel of Co_3O_4 and the calculated unit cell parameter was 8.0837 Å (Table 1), correlated well with the standard data (Co_3O_4, JCPDS: 42-1467). FTIR absorption bands observed at 558 and 654 cm^{-1} (Figure 1) were attributed to the stretching vibrations of Co-O, having the Co^{3+} in octahedral and tetrahedral coordination respectively, further confirmed the formation of Co_3O_4 spinel structure [16]. Nitrogen adsorption isotherm followed the type IV of IUPAC classification with H3 hysteresis loop (Figure 2), typically that of non-rigid aggregates of plate-like particles containing macrospores [17]. BET surface area and pore volume determined from N_2 adsorption were higher compared to values reported for Co_3O_4 obtained by different sol-gel method [18]. The percentage of micro, meso and macro pore volume were found to be 5.9, 82.1 and 12% respectively. SEM image of the cobalt oxide catalyst has shown the presence of aggregates Co_3O_4 (Figure 3), which further reinforces the conclusion from the N_2 adsorption characterization.

Table 1. Physicochemical properties of cobalt oxide catalyst

Unit cell parameters (Å)	B.E.T Surface area in m^2/g	Total Pore Volume in cm^3/g	Micro pore Volume in cm^3/g	Average pore diameter in Å
Cubic; a = 8.0837	413.63	0.555847	0.032666	50.887

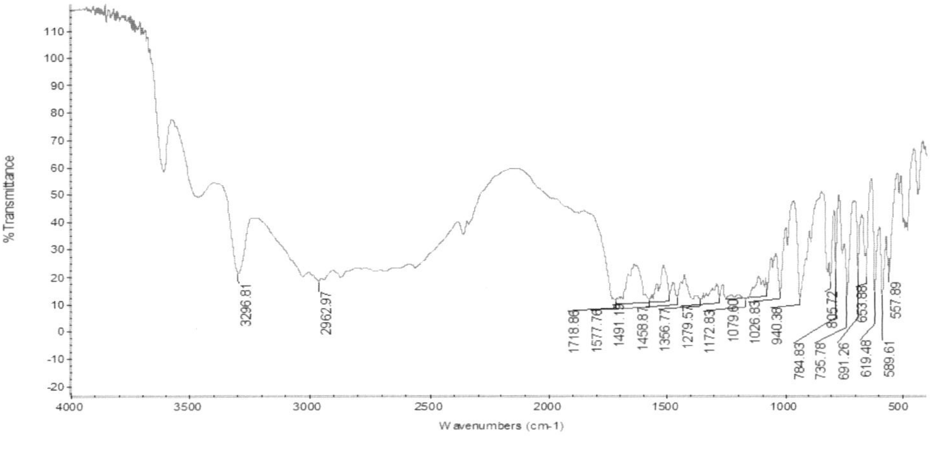

Figure 1. FTIR spectra of cobalt oxide catalyst. .

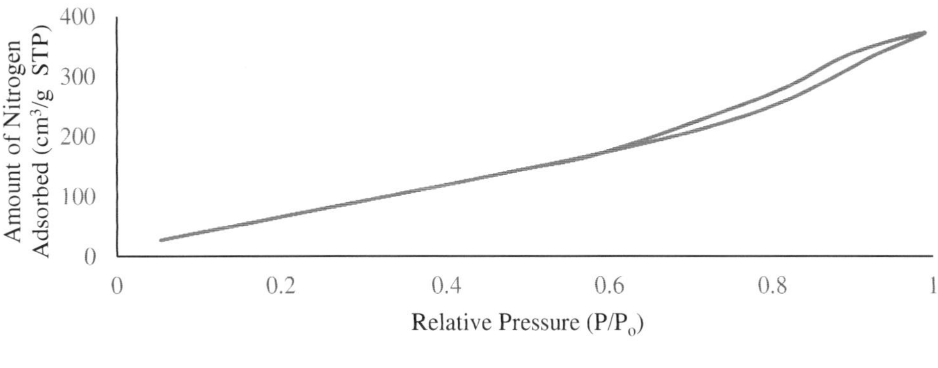

Figure 2. Nitrogen adsorption isotherm of cobalt oxide catalyst

Figure 3. SEM image of cobalt oxide catalyst

3.2. HTL of sugarcane bagasse

For all the catalytic runs, percentage of total oil yield was higher and the percentage conversion was lower compared to non-catalytic run (Figure 4). This was in accordance with the reported finding that the conversion of sugarcane was owing to the depolymerization of lignocellulose to WSO viz. xylose, mannose, glucose, galactose, p-coumaryl alcohol, coniferyl alcohol, sinapyl alcohol [12] catalyzed by OH^- and H^+ furnished by sub-critical water [19], independent of the externally added catalyst. Whereas the conversion of WSO to water immiscible oil was due to the deoxygenation activity of catalyst [12]. Percentage of total oil yield increased when the catalyst/biomass (C/B) weight ratio was increased from 0 to 0.4, mainly due to the increase in the yield of oil 2 and 3. Further increase to 0.6, resulted in the decrease of the total oil yield, due to the decomposition and re-polymerization of components of oil in to gas and solid residue respectively.

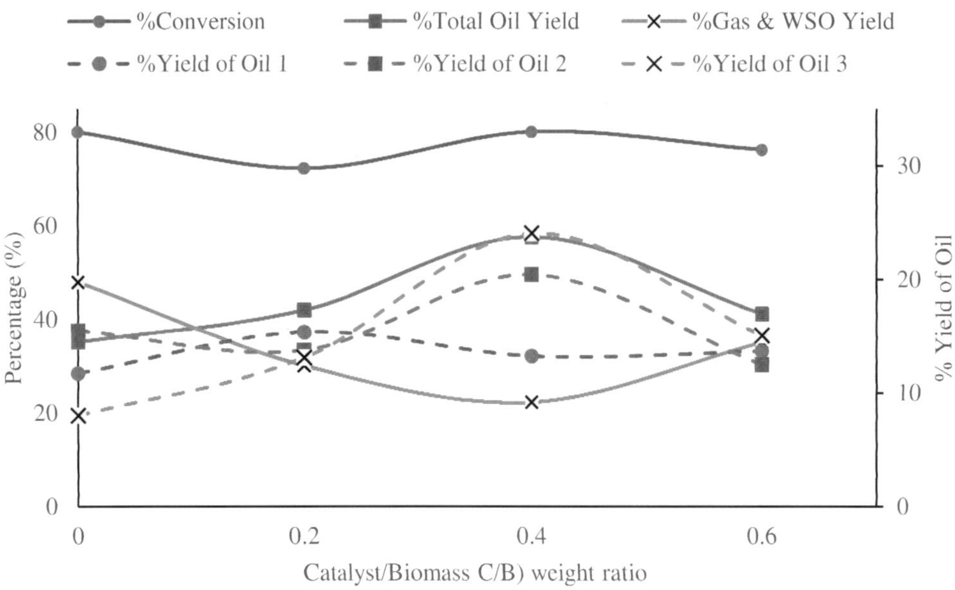

Figure 4. Influence of Catalyst/Biomass weight ratio on HTL of sugarcane bagasse

Percentage of conversion increased when the reaction time was increased from 0 to 30 min, remained constant up to 60 min and increased with the further hike in reaction time to 120 min (Figure 5). On the other hand, the percentage of total oil yield increased continuously with the increase in reaction time, despite no increase in the conversion from 30 to 60 min which was entirely due to the decrease in the percentage of gas & WSO yield. This indicated that the conversion of sugarcane bagasse by depolymerization of lignocellulose to WSO its subsequent deoxygenation to oil occurred consecutively. A similar observation was reported over Fe-Co oxide catalyst [12].

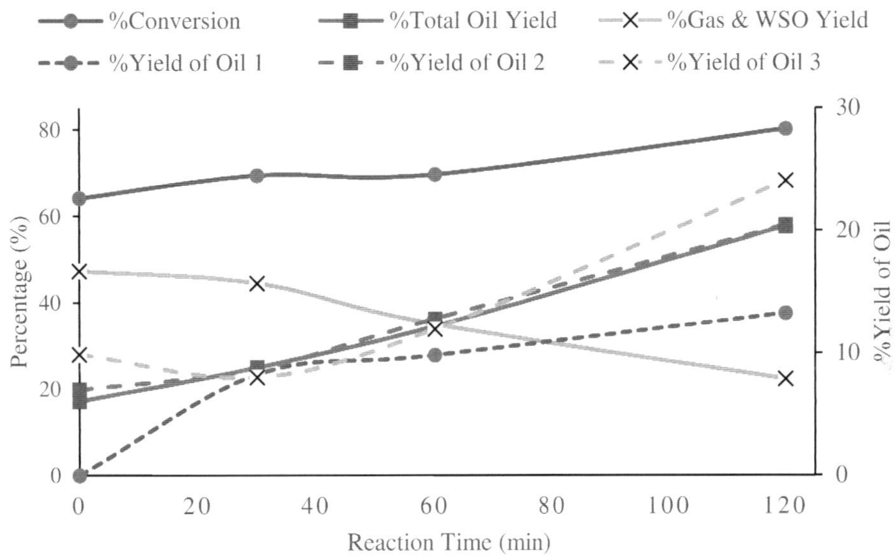

Figure 5. Influence of reaction time on HTL of sugarcane bagasse

Percentage of conversion as well as the total oil yield increased with increase in water/biomass ratio from 15 to 28 and then decreased with further increase to 40 (Figure 6). This reaffirmed the existence of the two consecutive chemical equilibrium, one between lignocellulose and WSO, and another between WSO and bio-oil. Increase in water/biomass ratio initially from 15 to 28 favored the conversion of sugarcane bagasse by the hydrolysis of lignocellulose to WSO and hence by its subsequent deoxygenation leading to higher oil yield. A further increase in the water/biomass ratio of 40 stabilized the WSO, thereby retarding its deoxygenation to bio-oil and hence the preceding step of hydrolysis of lignocellulose, accounting for the decrease in the percentage of conversion and total oil yield.

4. Conclusions

The spinel Cobalt oxide catalyst was found to have the deoxygenation activity required for the higher oil yield from the HTL of sugarcane bagasse. There was an optimum catalyst/biomass weight ratio of 0.4 giving a maximum oil yield, beyond that the catalyst helped only the undesirable decomposition and re-polymerization of oil. Two consecutive equilibrium was observed, one between lignocellulose and WSO another between WSO and oil. There was an optimum water/biomass ratio of 28 giving a maximum oil yield by shifting the equilibrium between lignocellulose and WSO towards WSO. Further increase to 40, retarded the equilibrium between WSO and oil by stabilizing the WSO, thus retarded the preceding equilibrium as well, and resulted in decrease of both conversion and oil yield.

Acknowledgements
The authors gratefully thank the Ministry of New and Renewable Energy, Government of India (Grant number F.No.7/184/2013-B F) for the financial support.

References

[1] Nel WP and Cooper CJ Implications of fossil fuel constraints on economic growth and global warming 2009 *Energy Policy* **37(1)** 166–80.

[2] Iyer GC, Edmonds JA, Fawcett AA, Hultman NE, Alsalam J, Asrar GR, Calvin KV, Clarke LE, Creason J, Jeong M, Kyle P, McFarland J, Mundra A, Patel P, Shi W and McJeon HC The contribution of Paris to limit global warming to 2 °C 2015 *Environ. Res. Lett.* **10(12)** 125002-11.

[3] Jeswani HK, Chilvers A and Azapagic A Environmental sustainability of biofuels: A review: Environmental sustainability of biofuels 2020 *Proc R Soc A Math Phys Eng Sci.* **476(2243)** 20200351-88.

[4] Lynd LR, Liang X, Biddy MJ, Allee A, Cai H, Foust T, Himmel ME, Laser MS, Wang M, Wyman CE Cellulosic ethanol: status and innovation 2017 *Curr. Opin. Biotechnol.* **45** 202–11.

[5] Zhou CH, Xia X, Lin CX, Tong DS and Beltramini J Catalytic conversion of lignocellulosic biomass to fine chemicals and fuels 2011 *Chem Soc Rev.* **40(11)** 5588–617.

[6] Axelsson L, Franzén M, Ostwald M, Berndes G, Lakshmi G, Ravindranath NH Perspective: Jatropha cultivation in southern India: Assessing farmers' experiences 2012 *Biofuels, Bioprod Biorefining.* **6(3)** 246–56.

[7] Seabra JEA and Macedo IC Comparative analysis for power generation and ethanol production from sugarcane residual biomass in Brazil 2011 *Energy Policy* **39(1)** 421–8.

[8] Soares LA, Rabelo CABS, Delforno TP, Silva EL and Varesche MBA Experimental design and syntrophic microbial pathways for biofuel production from sugarcane bagasse under thermophilic condition 2019 *Renew. Energy.***140** 852–61.

[9] Ahmed II and Gupta AK Sugarcane bagasse gasification: Global reaction mechanism of syngas evolution 2012 *Appl Energy* **91(1)** 75–81.

[10] Treedet W and Suntivarakorn R Design and operation of a low cost bio-oil fast pyrolysis from sugarcane bagasse on circulating fluidized bed reactor in a pilot plant 2018 *Fuel Process Technol.***179** 17–31.

[11] Singh R, Prakash A, Balagurumurthy B, Singh R, Saran S and Bhaskar T Hydrothermal liquefaction of agricultural and forest biomass residue: comparative study 2015 *J. Mater. Cycles Waste Manag.* **17(3)** 442–52.

[12] Govindasamy G, Sharma R and Subramanian S Effect of composition of iron-cobalt oxide catalyst and process parameters on the hydrothermal liquefaction of sugarcane bagasse 2020 *Bul.l Chem. React. Eng. Catal.***15(1)** 186–98.

[13] Akiya N and Savage PE Roles of water for chemical reactions in high-temperature water 2002 *Chem. Rev.* **102(8)** 2725–50.

[14] Soni VK, Dhara S, Krishnapriya R, Choudhary G, Sharma PR and Sharma RK Highly selective Co3O4/silica-alumina catalytic system for deoxygenation of triglyceride-based feedstock 2020 *Fuel* **266** 117065-72.

[15] Govindasamy G, Sharma R and Subramanian S Studies on the effect of heterogeneous catalysts on the hydrothermal liquefaction of sugarcane bagasse to low-oxygen-containing bio-oil 2019 *Biofuels* **10(5)** 665–75.

[16] Pudukudy M andYaakob Z. Sol-gel synthesis, characterisation, and photocatalytic activity of porous spinel Co_3O_4 nanosheets 2014 *Chem. pap.* **68(8)** 1087-96.

[17] Thommes M, Kaneko K, Neimark A V., Olivier JP, Rodriguez-Reinoso F, Rouquerol J and Sing KSW Physisorption of gases, with special reference to the evaluation of surface area and pore size distribution (IUPAC Technical Report) 2015 *Pure Appl. Chem.* **87(9–10)** 1051–69.

[18] Lin C, Ritter JA and Popov BN Characterization of Sol-Gel-Derived Cobalt Oxide Xerogels as Electrochemical Capacitors 1998 *J. Electrochem. Soc.* **145(12)** 4097–103.

[19] Toor S S, Rosendahl L and Rudolf A Hydrothermal liquefaction of biomass: A review of subcritical water technologies 2011 **36(5)** 2328–42.

IOP Conf. Series: Earth and Environmental Science 1042 (2022) 012008 doi:10.1088/1755-1315/1042/1/012008

Bio-oil by the hydrothermal liquefaction of municipal sewage sludge using hierarchical Fe-Co-ZSM-5 catalysts

Gopalakrishnan Govindasamy*, Praveen Kumar J, Ramatejaswi Pasupula
Department of Chemical Engineering, School of Engineering, University of Petroleum & Energy Studies(UPES), Dehradun-248007, India

*Email: gopalakrishnan@ddn.upes.ac.in

Abstract. Hydrothermal liquefaction (HTL), a mimic of natural formation of petroleum is the only thermochemical conversion having ability to convert the wet biomass to bio-oil, which upon hydrotreatment gives drop-in auto fuels. Yield of bio-oil, the energy product of HTL depends on the activity of the catalyst. Hence the synthesis of hierarchical Fe-Co-ZSM-5 was reported for the first time, the same was characterized and evaluated for the HTL of municipal sewage sludge to bio-oil. From their XRD characterization and calculated XRD crystallinity, 3 days was found to be the optimum crystallization time, as its crystallinity was 84% of hierarchical Al-ZSM-5 as standard, maximum among others crystallized at 2 and 4 days. From the characterization of Fe-Co-ZSM-5 obtained at 3 days by N_2 adsorption, mesoporosity was found to be 87.7% conforming its hierarchical nature and UV-visible reflectance spectrometry confirmed the incorporation of Fe and Co into the tetrahedral zeolite framework. Hydrothermal liquefaction of municipal sewage sludge studied at 250°C and initial H_2 pressure of 40 bar using the hierarchical Fe-Co-ZSM-5 (3 days) catalysts gave higher bio-oil yield of 23.8% compared to 10.26% for non-catalytic HTL.

1. Introduction

Population explosion, rapid urbanization, and increase in per capita consumption led to the steep increase in consumption of fossil fuels resulting in their fast depletion, increased CO_2 emission [1] and largescale generation of municipal liquid waste (MLW) and municipal solid waste (MSW) [2]. The MSW and MLW are inexhaustible and reliable sources of biomass available without any supply chain constrain. Thus developing and adopting waste to energy technologies to convert them to energy products helps to move from fossil fuel based economy to green and circular economy [3], to address both energy security and global warming. Municipal wastewater treatment plants generates energy rich sewage sludge (SS) of 0.2 to 0.3 tons for every million liters of wastewater treated by primary and secondary treatment [2]. Due to negative effects of disposal of SS by landfill and benefits of energy recovery from waste, anaerobic digestion (AD) [4] and hydrothermal liquefaction (HTL) [5,6] were reported for its conversion into biogas and bio-oil respectively as energy products. The reported retention time were 10-16 days and 17 min while the heating value of energy products per kg of SS processed were 14.85 and 39.5 MJ for AD and HTL respectively, indicated the swiftness and energy recovery efficiency of HTL. HTL carried out in hot compressed water is a mimic of natural process of petroleum formation [7] involving a three step mechanism. First step is depolymerisation of bio-polymers such as starch, lignocellulose, fats and proteins present in the biomass to sugars, organic acids, phenols and amino acids by acid-base catalyzed hydrolysis [8]. The second step is these hydrolysates which are water soluble organics (WSO) undergo secondary conversion such as isomerization, dehydration, decarboxylation, deamination leading to conversion of WSO to water immiscible bio-oil [9] . This step needs a catalyst having high deoxygenation activity and optimum reaction conditions to maximize the bio-oil yield [10]. The third step is undesirable repolymerization and cracking of components of oil to char and gas respectively.

Malins et al [6] investigated the water soluble Na_2CO_3, $FeSO_4$ and water insoluble Raney nickel and molybdenum sulphide catalysts for the HTL of SS at 300°C for 40 min and found that $FeSO_4$ gave the highest bio-oil yield of 45.58 wt.% while Na_2CO_3 gave the lowest bio-oil yield of 38.85 wt%. Prestigiacomo et al [11] evaluated $NiMo/Al_2O_3$, $CoMo/Al_2O_3$ and activated carbon felt catalysts, found that all the three gave low yield but high quality bio-crude with higher heating value compared to non-catalytic HTL. Shah et al [12] have found an increase in the bio-oil yield of 3 wt% with K_2CO_3 catalyst in comparison to non-catalytic HTL of SS owing to hydrolysis of carbohydrates by the cleavage of glycosidic bonds whereby reducing the formation of char. Gopalakrishnan et al [10] have studied Fe, Co, Mo and Ti oxides, their substituted MCM-41, and mixed metal oxides of Co-Mo, Fe-Mo, Ti-Zr, and Fe-Co as catalysts for hydrothermal liquefaction of sugarcane bagasse and reported that Fe-Co oxide yielded the highest bio-oil of 57.6 wt%. Bahri et al [13] reported an enhanced conversion of H_2 deficient and CO_2 containing syngas to transportation fuel over Fe-Co supported on hierarchical HZSM-5 prepared by post-synthesis modification. Steam assisted crystallization (SAC) of Al form of hierarchical ZSM-5 was reported using tetrapropylammonium hydroxide (TPAOH) as organic structure directing agent (OSDA) and found higher selectivity and catalyst life time for the conversion of methanol to aromatics [14]. To the best our knowledge there was no work reported on the SAC of hierarchical Fe-Co-ZSM-5. Hence the present study is aimed to synthesize iron, cobalt co-substituted hierarchical ZSM-5 by SAC, characterize and evaluate for hydrothermal liquefaction of SS to bio-oil.

2. Materials and methods

Ludox HS-40 and 35% tetraethylammonium hydroxide (TEAOH) solution were of Aldrich while ferric nitrate nonohydrate, cobaltous chloride hexahydrate, acetone, ethyl acetate and diethyl ether were of AR grade of SRL. The municipal inlet wastewater and mixed sewage sludge (primary/secondary sludge) were kindly provided by sewage treatment plant located at Kargi road, Dehradun, Uttarakhand, India.

2.1. Synthesis and characterization of hierarchical Fe-Co-ZSM-5

Fe-Co-ZSM-5 was synthesized using steam assisted crystallization (SAC) method. Calculated quantities of Ludox HS-40%, ferric nitrate, cobaltous chloride and 35% TEAOH solution as sources of silica, iron, cobalt and OSDA were mixed in to a gel of molar composition $0.0083Na_2O$ $0.0083(Fe_2O_3 + Co_2O_3)$ SiO_2 0.0767TEAOH $18.8H_2O$. The gel was heated at 80°C for 90 minutes till dry powder was obtained. The powder was taken in a Teflon cup and placed in a 300 ml autoclave. 12 ml of water was added into the space between Teflon cup and autoclave, closed and kept in air oven at 175°C. After the due crystallization time, autoclave was cooled to room temperature, filtered and filter cake was washed with distilled water till the pH of washings became 10. The filter cake was dried in an air-oven for 12 h at 110°C and calcined at 550°C for 6 hours. Three hierarchical Fe-Co-ZSM-5 catalysts were prepared using the above procedure at different crystallization time viz. 2, 3 and 4 days, labelled as ICZ 1, ICZ 2 and ICZ 3 respectively. Hierarchical Al-ZSM-5 was prepared by the same procedure except sodium aluminate replacing ferric nitrate and cobaltous chloride with 3 days crystallization time. Catalysts were characterized by X-ray diffraction using Bruker D8 Advance diffractometer with nickel-filtered CuKα radiation (λ = 1.5418 Å), in the 2θ range of 5–60°. The crystallinity was calculated by taking hierarchical Al-ZSM-5 as standard using the following equation;

$$\%\text{Crystallinity} = \frac{\Sigma \text{ Height of peaks at } 2\theta \text{ of } 7.9, 8.9, 23, 23.2 \text{ and } 23.9}{\Sigma \text{ Height of peaks at } 2\theta \text{ of } 7.9, 8.9, 23, 23.2 \text{ and } 23.9 \text{ of hierarchical Al}-\text{ZSM}-5} \quad (1)$$

The catalyst was studied by nitrogen adsorption at −195.85°C using micromeritics ASAP-2020 porosimeter with the built-in software for the determination of BET surface area and BJH pore-size distribution. Ultraviolet–visible reflectance spectra were recorded in the range of 190–800 nm with Jasco V-650 spectrometer using dehydrated $BaSO_4$ as the internal standard reference

2.2 Hydrothermal liquefaction of municipal sewage sludge

70.94 gm of sewage sludge containing 21.2% total solid (TS) was mixed with 29 ml of inlet wastewater (COD 300 mg/L) to get the feedstock of 100 ml containing 15% TS. Feed stock was transferred to high pressure stainless steel batch reactor (Figure 1) of volume 170 ml (Berghof BR 100), added with 1 g of catalyst under stirring and then kept over magnetic stirrer cum heater. After closing the reactor lid, it was purged with hydrogen to expel the air, pressurized with hydrogen up to 40 bar and then gas inlet was closed. The reactor was heated to 250°C and kept for 2 h, using temperature controller cum indicator. Then the reactor was cooled down to ambient temperature and the gaseous products were released. The rest of the product mixture was separated into Solid Residue (SR), Oil 1, Oil 2, Oil 3, and WSO by following the reported procedure [15]. %Yield of Total Oil, %WSO & Gas Yield and %Conversion were calculated by using the equations;

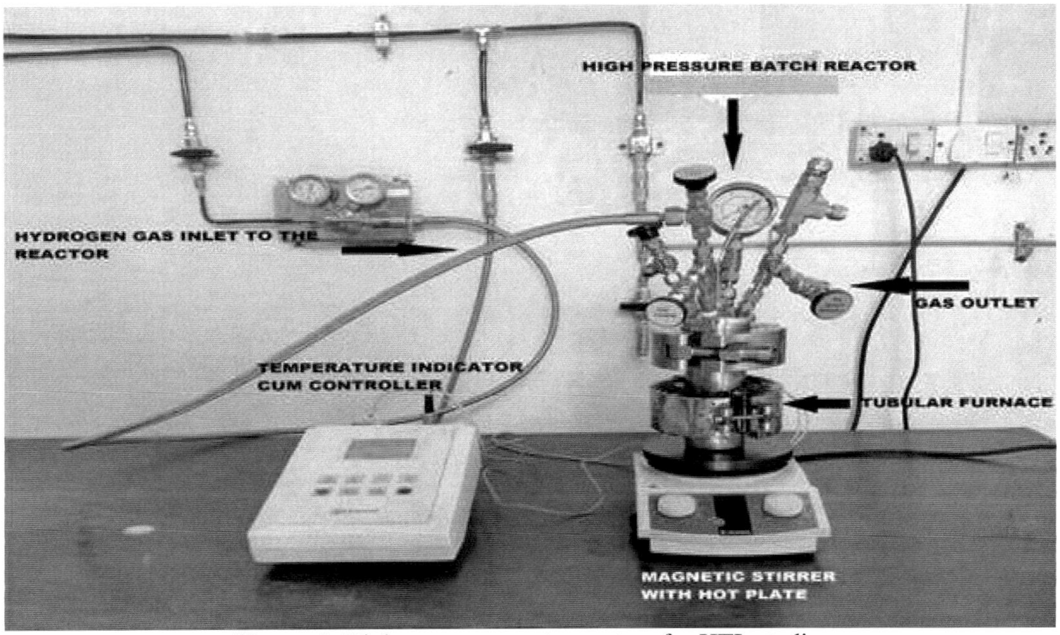

Figure 1. High pressure reactor system for HTL studies

$$\%\text{Yield of Total Oil} = \frac{\text{Yield of (Oil 1+Oil 2+Oil 3)}}{\text{TS of Feedstock}} \text{ X } 100 \quad (2)$$

$$\%\text{WSO \& Gas Yield} = \frac{\text{TS of Feedstock} - \text{Yield of (Oil 1+Oil 2+Oil 3+SR)}}{\text{TS of Feedstock}} \text{ X } 100 \quad (3)$$

$$\%\text{Conversion} = \frac{\text{TS of Feedstock} - \text{Yield of SR}}{\text{TS of Feedstock}} \text{ X } 100 \quad (4)$$

FTIR spectra of the oil samples were recorded using the Frontier model of Perkin Elmer.

3. Results and discussion

3.1. Catalyst characterization

The XRD pattern of the hierarchical Fe-Co-ZSM-5 synthesized at different crystallization time are presented in Figure 2 and their crystallinity in Table 1. XRD pattern of the hierarchical Fe-Co-ZSM-5 prepared at different crystallization time viz. 2, 3 and 4 days (Figure 2) have shown the characteristic peaks of hierarchical ZSM-5 [14]. From the Table 1, it could be inferred that that crystallinity increased from 2 to 3 days and no significant change observed with further increase to 4 days. Hence it may be concluded that 3 days is the optimum crystallization time for hierarchical Fe-Co-ZSM-5. Wang et al [16] reported a similar trend by varying the crystallization time from 0 to 24 h for SAC of hierarchical Al-ZSM-5, using tetraethylorthosilicate (TEOS) as silica source and TPAOH as OSDA, found that crystallinity increased from 0 to 12 h and did not change with further increase in crystallization time.

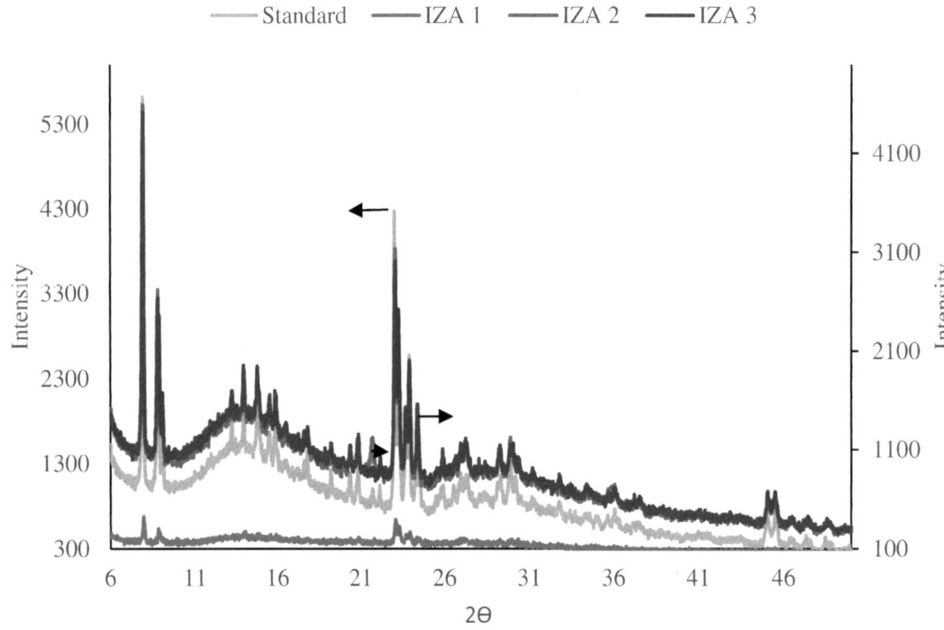

Figure 2. XRD pattern of hierarchical Fe-Co-ZSM-5 catalysts and Al-ZSM-5

Table 1. Crystallinity of hierarchical Fe-Co-ZSM-5 samples

S.No	Name of hierarchical Fe-Co-ZSM-5	Crystallization time (days)	%Crystallinity
1	ICZ 1	2	9.45
2	ICZ 2	3	84.00
3	ICZ 3	4	82.88

Nitrogen adsorption isotherm of ICZ 2 (Figure 3) followed type IVa of IUPAC classification of physisorption, reportedly characteristic of mesoporous materials with pores wider than 4 nm [17], conforming the formation of hierarchical Fe-Co-ZSM-5. The BET surface area and total pore volume (Table 2) were found to be similar to hierarchical Al-ZSM-5 reported by Jia et al [14], the mesopore volume was found to be higher compared to theirs.

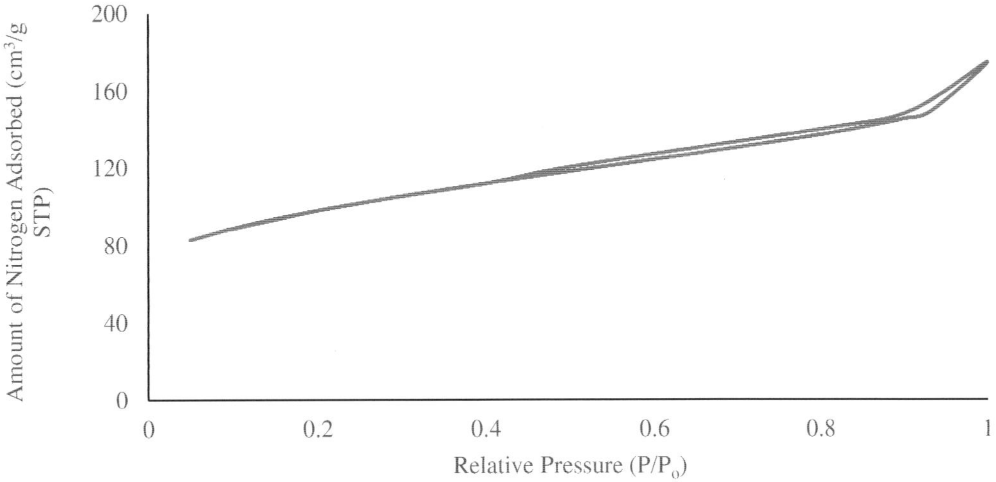

Figure 3. Nitrogen adsorption isotherm of ICZ 2

Table 2. Textural characteristics of ICZ 2

BET surface area (m^2/g)	Total pore volume (cm^3/g)	Micropore volume(cm^3/g)	Mesopore volume (cm^3/g)	%Mesoporosity $\left(\frac{\text{Mesopore volume}}{\text{Total porevolume}}\right)$ x 100
321.1	0.19	0.02	0.17	87.7

In the UV-visible reflectance spectra of ICZ 2 (Figure 4), peak observed around 220 nm could be attributed to tetrahedral Fe^{3+} of zeolite framework owing to the charge transfer $t_1 \rightarrow t_2$ and $t_1 \rightarrow e$ [18]. Peak at 250 nm attributed to the transfer of charge from the intra-framework oxygen to Co and another peak around 340 nm due to the interaction of Co with zeolite frame work oxygen [19], confirmed the isomorphic substitution Co^{2+} in to the zeolite framework.

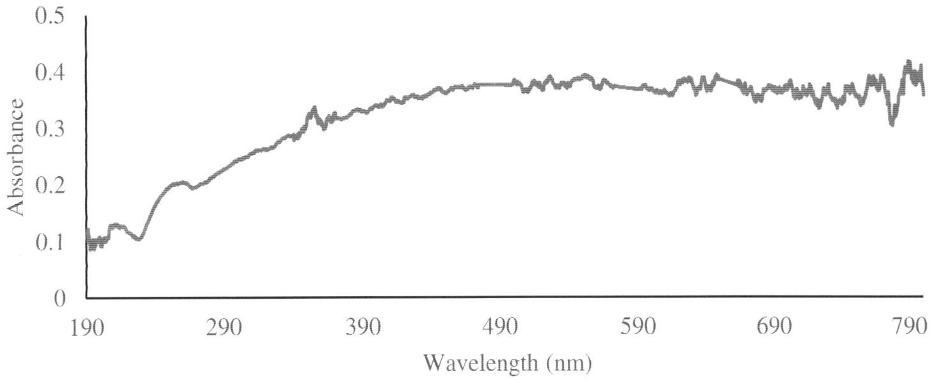

Figure 4. UV-Visible reflectance spectra of ICZ 2

IOP Conf. Series: Earth and Environmental Science 1042 (2022) 012008 doi:10.1088/1755-1315/1042/1/012008

3.2. Hydrothermal liquefaction studies

The product distribution obtained for the non-catalytic and catalytic hydrothermal liquefaction of municipal SS in the presence of two crystalline hierarchical Fe-Co-ZSM-5 catalysts viz. ICZ 2 and ICZ 3 are presented in Table 3. %Conversion was found to be similar for both catalytic and non-catalytic HTL confirmed the activity of both the catalysts to deoxygenate the WSO to bio-oil. Among the catalysts, hierarchical ICZ 2 has given 10.2 of higher %Total Oil Yield compared to ICZ 3 due to increase in the yield of all the three oils, owing to 1.1% higher in crystallinity. Lower oil yield of 23.8% compared to oil yield of 41% reported for HTL of SS [20] was possibly due to lower reaction temperature of 250°C employed in the present study compared to reaction temperature 347°C reported in the study under comparison.

Table 3. Product distribution of HTL of municipal sewage sludge

Catalyst	Product yield (g)				%Conversion	%Total oil yield	%WSO & gas yield
	SR	Oil 1	Oil 2	Oil 3			
Non-catalytic	2.70	0.49	0.53	0.52	82.0	10.26	71.71
ICZ 2	2.69	1.15	0.64	1.78	82.0	23.8	58.23
ICZ 3	2.52	0.42	0.55	1.07	83.2	13.6	69.6

From the FTIR spectra of oils (Figure 5), bands observed around 1640 and broad band around 3500 cm^{-1}, for all the three oils were due to the bending vibrations of C=C and stretching vibration of O-H respectively, indicated the presence of olefins and alcohols in all of them. A weak band observed for oil 1 at 2519 cm^{-1} was due to OH group attached to C=O group, indicated the presence of carboxylic acid. IR absorption band around 2075 cm^{-1} shown by oil 2 and 3 were due to C≡C vibration, thus indicated the presence of alkyne in them.

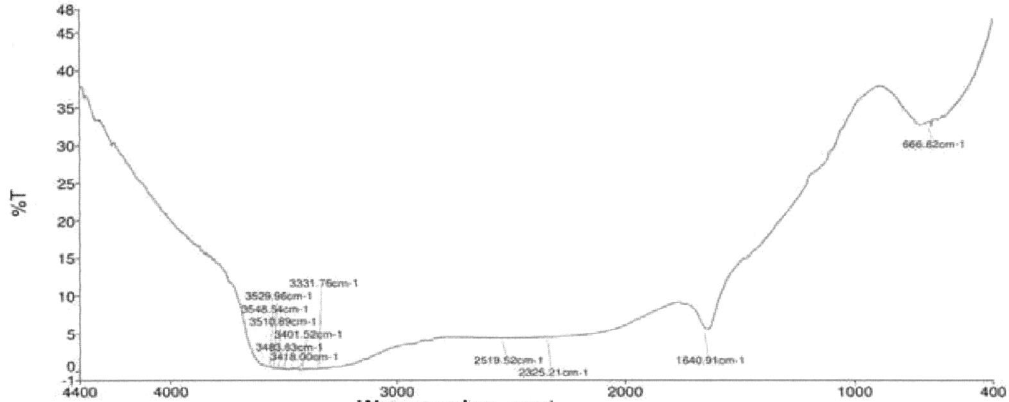

(a) Oil 1

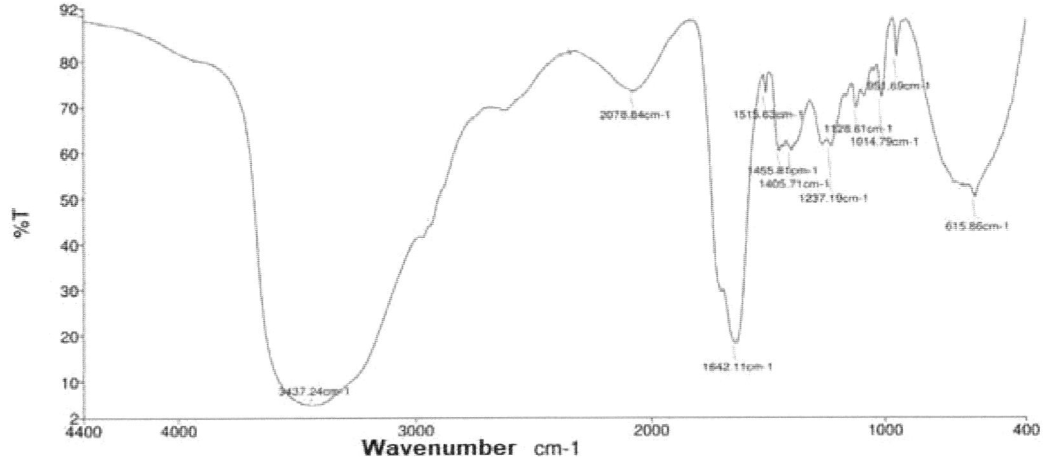

(b) Oil 2

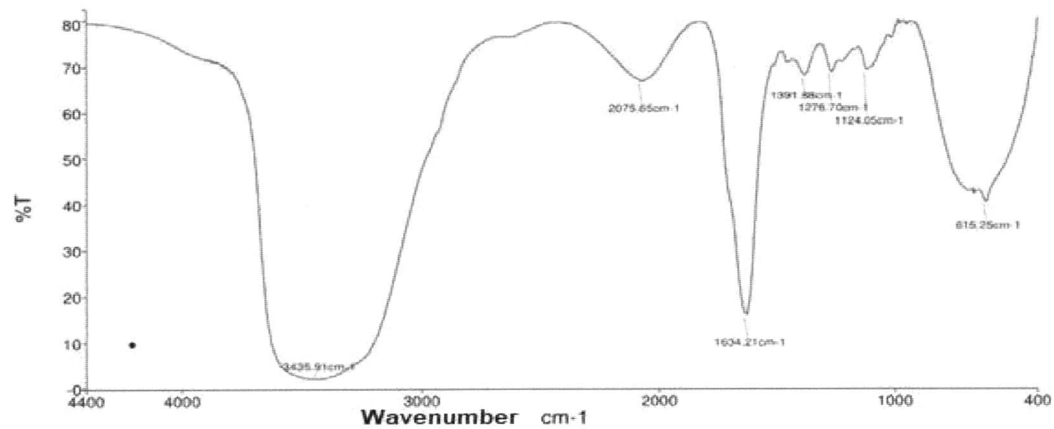

(c) Oil 3

Figure 5. FTIR spectra of oil samples

4. Conclusions

By steam assisted crystallization (SAC), hierarchical Fe-Co-ZSM-5 was first time synthesized, involving very low usage of TEAOH as OSDA, thus cost effective and green. From the XRD characterization, optimum crystallization time was found to be 3 days for the SAC of hierarchical Fe-Co-ZSM-5. The mesopore volume of Fe-Co-ZSM-5, synthesized with 3 days of crystallization time was 87.7% of the total pore volume, one of the highest reported so far, for hierarchical ZSM-5. From the evaluation of crystalline hierarchical Fe-Co-ZSM-5 synthesized at 3 and 4 days, for the HTL of municipal sewage sludge, both of them were found to yield more amount of bio-oil in comparison to non-catalytic HTL, thus confirmed their efficacy. The highest bio-oil yield of 23.8% was formed over hierarchical Fe-Co-ZSM-5 synthesized at 3 days of crystallization time due to its higher crystallinity, compared to 4 days. Optimization of reaction temperature and other process parameters will be the future study to increase the bio-oil yield further.

Acknowledgements
The authors acknowledge UPES R&D for funding this research work by SEED grant.

References

[1] Nel WP and Cooper CJ 2009 Implications of fossil fuel constraints on economic growth and global warming *Energy Policy* **37(1)** 166–80.

[2] Seiple TE, Coleman AM and Skaggs RL 2017 Municipal wastewater sludge as a sustainable bioresource in the United States *J. Environ. Manage.* **197** 673–80.

[3] Cucchiella F, D'Adamo I and Gastaldi M 2019 Sustainable Italian cities: The added value of biomethane from organic waste *Appl. Sci.* **9(11)** 2221-48.

[4] Bhatt AH and Tao L 2020 Economic perspectives of biogas production via anaerobic digestion *Bioengineering* **7(3)** 74–92.

[5] Marrone PA, Elliott DC, Billing JM, Hallen RT, Hart TR, Kadota P, Moeller JC, Randel MA and Schmidt AJ 2018 Bench-Scale Evaluation of Hydrothermal Processing Technology for Conversion of Wastewater Solids to Fuels *Water Environ. Res.* **90(4)** 329–42.

[6] Malins K, Kampars V, Brinks J, Neibolte I, Murnieks R and Kampare R 2015 Bio-oil from thermo-chemical hydro-liquefaction of wet sewage sludge *Bioresour. Technol.* **187** 23–9.

[7] Gollakota ARK, Kishore N and Gu S 2018 A review on hydrothermal liquefaction of biomass *Renew. Sustain. Energy. Rev.* **81** 1378–92.

[8] Toor SS, Rosendahl L and Rudolf A 2011 Hydrothermal liquefaction of biomass: A review of subcritical water technologies *Energy* **36(5)** 2328–42.

[9] Zhang X, Wilson K and Lee AF 2016 Heterogeneously Catalyzed Hydrothermal Processing of C5-C6 Sugars *Chem. Rev.* **116(19)** 12328–68.

[10] Govindasamy G, Sharma R and Subramanian S 2019 Studies on the effect of heterogeneous catalysts on the hydrothermal liquefaction of sugarcane bagasse to low-oxygen-containing bio-oil *Biofuels* **10(5)** 665–75.

[11] Prestigiacomo C, Costa P, Pinto F, Schiavo B, Siragusa A, Scialdone O and Galia A 2019 Sewage sludge as cheap alternative to microalgae as feedstock of catalytic hydrothermal liquefaction processes *J. Supercrit. Fluids* **143** 251–8.

[12] Shah AA, Toor SS, Seehar TH, Nielsen RS, Nielsen AH, Pedersen TH and Rosendhal LA 2020 Bio-crude production through aqueous phase recycling of hydrothermal liquefaction of sewage sludge. *Energies* **13(2)** 493-510.

[13] Bahri S, Patra T, Sonal and Upadhyayula S 2019 Synergistic effect of bifunctional mesoporous ZSM-5 supported Fe-Co catalyst for selective conversion of syngas with low Ribblet ratio into synthetic fuel *Microporous Mesoporous Mater.* **275** 1–13.

[14] Jia Y, Wang J, Zhang K, Chen G, Yang Y, Liu S, Ding C, Meng Y and Liu P 2018 Hierarchical ZSM-5 zeolite synthesized via dry gel conversion-steam assisted crystallization process and its application in aromatization of methanol *Powder Technol.* **328** 415–29.

[15] Govindasamy G, Sharma R and Subramanian S 2020 Effect of composition of iron-cobalt oxide catalyst and process parameters on the hydrothermal liquefaction of sugarcane bagasse *Bull. Chem. React. Eng. Catal.* **15(1)** 186–98.

[16] Wang Y, Song J, Baxter NC, Kuo GT and Wang S 2017 Synthesis of hierarchical ZSM-5 zeolites by solid-state crystallization and their catalytic properties *J Catal.* **349** 53–65.

[17] Thommes M, Kaneko K, Neimark AV, Olivier JP, Rodriguez-Reinoso F, Rouquerol J and Sing KSW 2015 Physisorption of gases, with special reference to the evaluation of surface area and pore size distribution (IUPAC Technical Report) *Pure Appl. Chem.* **87(9–10)** 1051–69.

[18] Liu S, Ren J, Zhu S, Zhang H, Lv E, Xu J and Li YW 2015 Synthesis and characterization of Fe-substituted ZSM-22 zeolite catalyst with high n-dodecane isomerization performance *J. Catal.* **330** 485–96.

[19] Zhang N, Shi L, Yu T, Li T, Hua W and Lin C 2015 Synthesis and characterization of pure STW-zeotype germanosilicate , Cu- and Co-substituted STW-zeotype materials *J. Solid State Chem.* **225** 271–7.

[20] Snowden-swan L, Seiple T, Fox S, Radovcich S, Mathias P and Taylor M. 2021 Wet Waste Hydrothermal Liquefaction and Biocrude Upgrading to Hydrocarbon Fuels : 2020 State of Technology.

Environmentally balanced preconditions for use of logging residues in forests of Ukrainian Carpathians for energy production

R Vasylyshyn[1*] Lakyda [1], O Soshenskyi[1], M Lakyda[1], O Melnyk[2], V Slusarchuk[3]

[1]National University of Life and Environmental Sciences of Ukraine, Heroiv Oborony str., 15, Kyiv, Ukraine, 03041
[2]Boiarka Forest Research Station, Lisodoslidna str., 12, Boiarka, Kyiv region, Ukraine, 08150
[3]Storozhynets Forestry College, Kreitera str., 1, Storozhynets, Chernivtsi region, Ukraine, 59000

[*]Email: r.vasylyshyn@nubip.edu.ua

Abstract. Use of woody biomass of logging residues at plots where different types of logging has occurred as a planned management activity has to be based on the analysis of relevant environmental risks. These risks include biodiversity losses, soil fertility reduction, and formation of dangerous volumes of flammable materials in forests during the fire-hazardous period, which is especially relevant on plots with high recreational load. As a result of the research, we propose indicative utilization rates for logging residues at sites where care or final cuts occur in spruce, beech and fir dominated stands in various growth conditions. The information basis of the research is formed by the research data on distribution of the relevant forest stands in different types of forest growth conditions, as well as on influence of forest growth conditions on processes of biological destruction and accumulation of forest flammable materials at plots with intensive recreational load. The proposed utilization rates for logging residues serve as an information component in the process of formation of regional strategy of forest bioenergy development.

1. Introduction

In the Ukrainian Carpathians, logging residues act as an important component of the overall energy potential of woody biomass in the region. At the same time, the priority of environmental functions of mountain forests determines application of environmentally oriented approaches to logging. In this context, environmental restrictions, which minimize negative environmental impact of use of logging residues, becomes especially relevant when assessing their biomass potential [1].

The key ecological precondition for energy use of logging residues is to ensure the biological cycling of nutrients in the system "forest ↔ soil". To this end, there is a need to form indicative rates for the amount of logging residues that have to be left on logging sites for biodegradation [2]. At the same time, another relevant issue is linked with providing a certain level of fire safety in forests with a significant anthropogenic load due to the possible accumulation of significant amounts of forest combustible materials [3].

2. Analysis of recent research and publications

The forests of Ukrainian Carpathians, having a high level of ecological vulnerability, are a quantitative and qualitative criterion of ecological balance in the region's natural environment. They therefore require special attention in matters of restoration, conservation, and rational use.

In Ukraine, active research on use of forest biomass for energy purposes is due to the need to form an energy-independent strategy for the national economy's development [4]. At the same time, researchers often do not take into account the environmental role and peculiarities of functioning of forest plant communities when estimating potential volumes of woody biomass (including logging residues) within areas covered by different types of cuts [5, 6]. This approach does not take into account the cyclical need of forest soils for mineral nutrients intake. This kind of need may be secured as a result of biodegradation of forest organic residues.

The research published in scientific literature [7] states that in coniferous and deciduous trees, the largest amount of mineral nutrients is contained in the foliage fraction of live biomass (leaves and needles). According to this indicator, foliage significantly outnumbers the fraction of crown branches. At the same time, according to the results of this study, the average content of mineral nutrients in branches up to 3.0 cm in diameter is higher than in branches with a diameter of 3.1-6.0 cm. In particular, nitrogen content in birch branches up to 3.0 cm equals 0.9 %, whereas in those between 3.1 and 6.0 cm in diameter – only 0.34 %. Similar ratios have also been recorded for potassium, magnesium, and phosphorus. The scientists, however, have not proposed any specific utilization rates for logging residues as an alternative energy source.

The ecological consequences for forest plant communities resulting from energy use of wood biomass in Swedish forests have also been studied by scientists [8], who note that increasing wood biomass harvesting may have negative consequences for the environment, fertility and chemical composition of forest soils, and from the point of view of biodiversity conservation. At the same time, the research sets the limit of 2.5 times maximal increase for possible use of logging residues, which will not violate the principles of sustainable forest management. A similar study was carried out by Polish scientists [9], who analyzed the physical and chemical properties of chipped forest residues in pine stands to develop information support of solid biofuel production.

Wood harvesting which includes removal of both trunks and crown components (foliage and all the branches) can lead to the loss of a significant amount of nutrients in the soil and its oxidation, which affects the productivity of subsequently planted stands on the same plot [10]. These environmental risks of logging residues and stumps use are the subject of research in many countries [11–16], as they act as a factor that influences the introduction of sustainable forest management.

Use of logging residues for energy purposes is also a prerequisite for low-carbon industrial production, however, according to the scientific publications [17], this effect is long-term by its nature. In the short term, this process guarantees a sustainable reduction in greenhouse gas emissions.

An important aspect of research in this direction is to predict the impact of use of logging residues on the formation of forest ecosystems' biodiversity and the efficiency of their functions. In particular, a significant use of logging residues can have negative impact on the formation of mammalian populations and water quality [18]. Under certain conditions, when increasing the volume of wood biomass harvesting, a significant probability of negative impact on forest ecosystems, their biodiversity and ability to contribute to fundamental ecosystem services arises [19]. However, provided a balanced consumption of this resource, forest bioenergy may become an important component of global climate change mitigation measures. This approach also enables reduction of concentration of forest combustible materials in the form of dead wood [20] in forest areas, and, accordingly, reduces fire hazard in forests with significant anthropogenic load.

The presented analysis of scientific publications allows to conclude that the use of logging residues requires an environmentally balanced approach, which can mitigate the possible negative consequences of excessive use of this resource for forest ecosystems, forest biodiversity, and ecosystem functions, which is especially important in mountainous regions.

3. The purpose of this work

The purpose of this scientific work is to establish indicative utilization rates for logging residues biomass at logging areas in compliance with the environmentally balanced approach to the nutrients cycle and ensuring a low fire hazard level.

4. Materials and methods of research

This research covers forested areas within the Ukrainian Carpathians on the territory of more than 2 million hectares. The region is dominated by spruce, beech and fir stands, whose total share is about 70 % of the area of forests. Mid-aged stands created in the middle of last century are the most widely represented, they account for more than 50 % of forested area. The share of young, maturing and mature stands is at the level of 15-17 %.

The key principle for forming quantitative values of utilization rates for logging residues consists in consideration of three basic factors of influence: soil fertility (4 classes, A – infertile conditions, B – fairly infertile conditions, C – fairly fertile conditions, D – fertile conditions) – determines the relative soil fertility (used to assess the necessity to increase the amount of nutrients in forest soils), soil moisture (6 classes, 0 – very dry, 1 – dry, 2 – fresh, 3 – moist, 4 – damp, and 5 – wet) – determines soil moisture (to assess the potential intensity of biodegradation and humification of logging residues), and national scale fire hazard class (to predict the consequences of forest fires in areas where forest combustible materials accumulate). Table 1 presents the share of the forest types (types of forest growth conditions) within the studied forest areas.

Table 1. Distribution of area covered by stands of dominant tree species in the study region by types of forest growth conditions.

Soil fertility class	Area of forest stands, thou. ha				
	total	incl. by soil moisture classes			
		fresh (2)	moist (3)	damp (4)	wet (5)
		Norway spruce stands			
Infertile *(A)*	0.53	–	0.53	–	–
Fairly infertile *(B)*	28.64	0.29	27.98	0.37	–
Fairly fertile *(C)*	468.46	3.41	461.76	3.30	–
Fertile *(D)*	93.94	2.09	91.70	0.15	–
Total	591.57	5,79	581.96	3.82	–
		European beech stands			
Fairly infertile *(B)*	0.42	0.22	0.20	–	–
Fairly fertile *(C)*	272.78	44.54	228.12	0.12	–
Fertile *(D)*	401.55	113.50	288.01	0.02	0.02
Total	674.75	158.26	516.33	0.15	0.02
		Silver fir stands			
Fairly infertile *(B)*	0.09	–	0.09	–	–
Fairly fertile *(C)*	78.10	1.41	75.99	0.70	–
Fertile *(D)*	51.52	0.40	50.94	0.17	–
Total	129.71	1.81	127.03	0.87	–

To formulate the utilization rates for logging residues in stands dominated by the main forest-forming species of Ukrainian Carpathians, the residues were differentiated into two categories depending on their thickness: up to 1 cm, and more than 1 cm. Within this research, we did not consider stumps as a potential raw material for energy use because they play an important soil protective role in mountain conditions.

The final values of indicative utilization rates for logging residues are presented separately for care cuts and final cuts in percent with an accuracy of 10 % aiming at ease of practical application.

5. Research results

The annual volume of timber logging in the research region is about 2.6 million m^3. In the process of harvesting, more than 400 thou. m^3 of logging residues are formed annually. Those are mainly crown branches and parts of tree trunks. A certain part of this resource may serve as an important source of renewable energy. Currently, only 10 to 15 % of logging residues are used for energy production purposes.

The current management practice in the region outlines leaving more than 90 % of logging residues for biodegradation in the areas covered by care cuts. Given the availability of a necessary production capacity, it is quite possible to use a part of this resource for production of thermal energy or solid biofuels (wood chips, pellets, briquettes, etc.) without harming the environment.

Table 2 contains the proposed indicative utilization rates for logging residues in areas where such management activities as care and tending cuts take place.

Table 2. Utilization rates for logging residues at care cuts.

Logging residues fraction	Purpose of use of logging residues	Index of type of forest growth conditions	Indicative utilization rate for logging residues, %
	Norway spruce and silver fir stands		
Branches with foliage, $d < 1$ cm*	for biodegradation	all types	100
Branches, $d > 1$ cm	for heat energy production and solid biofuels production	A_3	30
		B_2–B_4	40
		C_2–C_4, D_4	50
		D_2–D_3	60
	European beech stands		
Branches with foliage, $d < 1$ cm*	for biodegradation	all types	100
Branches, $d > 1$ cm	for heat energy production and solid biofuels production	B_2–B_3	50
		C_2–C_3, D_2–D_3	60
		C_4, D_4	40

*Note. Branches from fraction of tree greenery.

As a result of the research, we recommend leaving for biodegradation all the branches up to 1 cm in diameter, which are part of the fraction of tree greenery (fine branches with foliage).

Given that in deciduous stands the process of foliage biodegradation is more intense compared to coniferous stands, the utilization rates for European beech stands are 10 % higher in the respective types of forest growth conditions than for Norway spruce and silver fir stands due to the peculiarities specific for the coniferous and hardwood broadleaved tree species groups [21]. Also, in local areas that are subject to a significant recreational load during fire hazard period, we propose to increase the utilization rates by 10–20%.

The logging residues utilization rates at final cuts are higher in comparison with those for care and tending cuts. This is due to the fact of accumulation of a much larger stock of mineral nutrients in mature stands compared to young and mid-aged ones (Table 3). Under these utilization rates, the shelterwood silvicultural system will be capable to provide an effective natural regeneration of forests within the research region.

Table 3. Utilization rates for logging residues at final cuts.

Logging residues fraction	Purpose of use of logging residues	Index of type of forest growth conditions	Indicative utilization rate for logging residues, %
Norway spruce and silver fir stands			
Branches with foliage, $d < 1$ cm*	for biodegradation	all types	100
Branches, $d > 1$ cm	for heat energy production and solid biofuels production	A_3	40
		B_2–B_4	50
		C_2–C_4, D_4	60
		D_2–D_3	70
European beech stands			
Branches with foliage, $d < 1$ cm*	for biodegradation	all types	100
Branches, $d > 1$ cm	for heat energy production and solid biofuels production	B_2–B_3	60
		C_2–C_3, D_2–D_3	70
		C_4, D_4	50

*Note. Branches from fraction of tree greenery.

Making use of the presented indicative utilization rates for logging residues use may allow to draw in as much as 150 thou. m^3 of energy feedstock to the region's energy balance, while securing biodiversity conservation requirements, proper nutrients cycling in the in the system "forest ↔ soil", and minimization of forest fires during fire hazard seasons.

6. Conclusions

This research contributes to solving the problem of complex, balanced and environmentally safe use of forest resources in the region by proposing indicative utilization rates for logging residues in Norway spruce, silver fir and European beech stands. Application of the utilization rates enables extended inclusion of logging residues to the region's energy balance without any significant damage to forest ecosystems functioning. Utilization of 30 to 60 % of the amount of logging residues from care cuts and 40 to 70 % – from final cuts may help draw ca. 150 thou. m^3 of woody biomass for heat production. Use of the proposed rates in forest management planning serves as a tool for formation of a regional strategy for forest bioenergy development.

References

[1] Vasylyshyn R 2017 Theoretical and methodological basis for assessment of energy potential of forests' woody biomass on the sustainable forest management basis *Scientific herald of Forestry Academy of Sciences of Ukraine* **15** pp 82–89 doi.org/10.15421/411710

[2] Vasylyshyn R 2018 *Environmental and energy potential of forests in Ukrainian Carpathians and its sustainable use* (Kyiv: LLC Komprint) p 305

[3] Soshenskyi O, Zibtsev S, Gumeniuk V, Goldammer J G, Vasylyshyn R and Blyshchyk V 2021 The current landscape fire management in Ukraine and strategy for its improvement *Environmental & Socio-economic Studies* **9 (2)** pp 39–51

[4] Geletukha G and Zheliezna T 2021 Prospects for Bioenergy Development in Ukraine: Roadmap until 2050 *Ecological Engineering & Environmental Technology* **22 (5)** pp 73–81 doi.org/10.12912/27197050/139346

[5] Geletukha G, Zheliezna T, Pastukh A and Drahniev S 2018 *Possibilities for harvesting wood fuel in Ukraine's forests* (Kyiv: BAU) p 29

[6] Prokip A 2011 *Securing energy safety: past, present and future* (Lviv: EUCC) p 154

[7] Shatravko V 2013 Research of mineral elements in logging residues in stands of the main forest forming tree species of Belarus *Publications of the Belarus State Technological University* **21** pp 117–119

[8] De Jong J, Akselsson C, Egnell G, Lofgren S and Olsson B 2017 Realizing the energy potential of forest biomass in Sweden – How much is environmentally sustainable? *Forest Ecology and Management* **383** pp 3–16

[9] Nurek T, Gendek A and Roman K 2019 Forest Residues as a Renewable Source of Energy: Elemental Composition and Physical Properties *Bioresources* **14 (1)** pp 6–20

[10] Achat D L, Deleuze C, Landmann G, Pousse N, Ranger J and Augusto L 2015 Quantifying consequences of removing harvesting residues on forest soils and tree growth – A meta-analysis. *Forest Ecology and Management* **348** pp 124–141

[11] Giuntoli J, Caserini S, Marelli L, Baxter D and Agostini A 2015 Domestic heating from forest logging residues: environmental risks and benefits *Journal of Cleaner Production* **99** pp 206–216 doi.org/10.1016/j.jclepro.2015.03.025

[12] Tarvainen O, Hekkala A M, Kubin E, Tamminen P, Murto T and Tolvanen A 2015 Soil disturbance and early vegetation response to varying intensity of energy wood harvest *Forest Ecology and Management* **348** pp 153–163 doi.org/10.1016/j.foreco.2015.04.001

[13] Tarvainen O, Saravesi K, Pennanen T, Markkola A M, Suokas M and Fritze H 2020 Fungal communities in decomposing wood along an energy wood harvest gradient *Forest Ecology and Management* **465** 118070 doi.org/10.1016/j.foreco.2020.118070

[14] Kaarakka L, Vaittinen J, Marjanen M, Hellsten S, Kukkola M, Saarsalmi A and Helmisaari H S 2018 Stump harvesting in Picea abies stands: Soil surface disturbance and biomass distribution of the harvested stumps and roots *Forest Ecology and Management* **425** pp 27–34 doi.org/10.1016/j.foreco.2018.05.032

[15] Maillard F, Leduc V, Bach C, Reichard A, Fauchery L, Saint-Andre L and Buee M 2019 Soil microbial functions are affected by organic matter removal in temperate deciduous forest *Soil Biology & Biochemistry* **133** pp 28–36

[16] Ojanen P, Makiranta P, Penttila T and Minkkinen K 2017 Do logging residue piles trigger extra decomposition of soil organic matter? *Forest Ecology and Management* **405** pp 367–380 doi.org/10.1016/j.foreco.2017.09.055

[17] Repo A, Bottcher H, Kindermann G and Liski J 2015 Sustainability of forest bioenergy in Europe: land-use-related carbon dioxide emissions of forest harvest residues *Global Change Biology Bioenergy* **7(4)** pp 877–887 doi.org/10.1111/gcbb.12179

[18] Ranius T, Hamalainen A, Egnell G, Olsson B, Eklof K, Stendahl J and Felton A 2018 The effects of logging residue extraction for energy on ecosystem services and biodiversity: A synthesis *Journal of Environmental Management* **209** pp 409–425

[19] Eggers J, Melin Y, Lundstrom J, Bergstrom D and Ohman K 2020 Management Strategies for Wood Fuel Harvesting-Trade-Offs with Biodiversity and Forest Ecosystem Services *Sustainability* **12(10)** 4089 doi.org/10.3390/su12104089

[20] Forsius M, Akujarvi A, Mattsson T, Holmberg M, Punttila P, Posch M and Vihervaara P 2016 Modelling impacts of forest bioenergy use on ecosystem sustainability: Lammi LTER region, southern Finland *Ecological Indicators* **65** pp 66–75 doi.org/10.1016/j.ecolind.2015.11.032

[21] Sytnyk S, Lovynska V and Lakyda I 2017 Foliage biomass qualitative indices of selected forest forming tree species in Ukrainian steppe *Folia Oecologica* **44(1)** pp 38-45 doi.org/10.1515/foecol-2017-0005

IOP Conf. Series: Earth and Environmental Science 1042 (2022) 012010 doi:10.1088/1755-1315/1042/1/012010

Energy potential of woody biomass in Ukraine's forests and prospects for its utilization as an alternative energy source

R Vasylyshyn[1*] I Lakyda[1], Yu Yurchuk[2], M Lakyda[1], O Melnyk[3], R Bondarchuk[4]

[1]National University of Life and Environmental Sciences of Ukraine, Heroiv Oborony str., 15, Kiyv, Ukraine, 03041
[2]Popilnia state forestry enterprise, Bohdan Khmelnytskyi str., 61, Popilnia, Zhytomyr region, Ukraine, 13501
[3]Boiarka Forest Research Station, Lisodoslidna str., 12, Boiarka, Kyiv region, Ukraine, 08150
[4]Ovruch specialized state forestry enterprise, Zhytomyrska str., 35, Ovruch, Zhytomyr region, Ukraine, 11100

*Email: R.Vasylyshyn@nubip.edu.ua

Abstract. Woody biomass in Ukraine is currently seen as an important resource for securing development of national renewable energy sector, avoiding dependence on the imported fossil fuels, and for building a low-carbon economy. This research highlights the indices of the overall energy potential of Ukraine's forests of the dominant tree species, expressed in units of primary energy. The indices are differentiated by biomass components and stand age groups. The information basis of the research consists of two main components: data on biometric assessment of trees collected at temporary sample plots, and mathematical models of biomass expansion factors. The research features the assessment results on the annual potential of woody biomass for stem wood (fuel wood), logging residues and wood processing residues. This resource forms a part of the sources of renewable energy and can be used for securing energy needs of the society. The proposed indices of woody biomass energy potential serve as an information basis for shaping the regional programs of forest bioenergy development in Ukraine.

1. Introduction

Development of renewable energy production forms one of the main directions for decarbonization of Ukraine's energy sector. This is reflected in the Strategy for low-carbon development, approved by the Cabinet of Ministers of Ukraine back in 2018. This document sets aims for increased carbon sequestration and decrease in greenhouse gases emissions in the forestry sector. These aims are achieved through formation of an efficient forest fire prevention system [1], and dynamic development of forest bioenergy sector [2].

Under the current circumstances, development of energy sector in Ukraine is perceived through amounts and costs of imported energy producing materials, which dominate the country's energy balance. Aiming at moving away from energy dependence of the national economy and improving the condition of the environment, research on woody biomass energy potential has been initiated. Biomass is currently seen as a strategic energy resource countrywide, whose utilization can also help reduce environmental stress through reducing emissions of harmful substances originating from fossil fuels burning. This is especially relevant considering that the search for alternative sources of renewable energy under conditions of a complete energy dependence of Ukraine on energy exporting countries, is a strategic task for most sectors of the national economy, including forestry [3].

2. Analysis of recent research and publication

The generally accepted practice of assessing the energy function of forests in the world has two main directions, according to which its economic (1) or biophysical (2) assessment is prioritized. The second direction, historically, has developed significantly in Ukraine. It is largely based on forest biological productivity research, which provide a detailed scientific study of questions at the local and regional scale [4, 5, 6] in line with the IPCC's (Intergovernmental Panel on Climate Change) current vector. A considerable attention in the course of these scientific developments is paid to the quantification of individual ecosystem functions of forests [7].

The worldwide research in this area, among others, considers the issue of forecasting energy potential in countries with domestic markets that do not affect international markets. For example, in the case of Colombia, energy potential is assessed and forecasted using an optimization scenario that maximizes annual profits from local production and imports in the country, taking into account a set of constraints, as well as demographic, macroeconomic and market data [8]. Forest plant communities can serve as a source of biomass, which features high energy potential and capable of replacing fossil fuels. Currently, there are studies that analyze the spatial distribution of biomass and the location of storage centers in southwestern Europe by quantifying, mapping, and defining biomass taking into account environmental and economic aspects [9].

According to the current understanding of the future development of mankind, timber harvesting for bioenergy purposes is likely to increase in the future to replace fossil fuels, especially in countries with significant forested areas. However, the adverse environmental impact of increased biomass harvesting can be significant over long periods of time, in particular on soil, water, biodiversity and productivity. The results of research carried out in Sweden indicate the possibility of 2.5-fold increase of harvest of logging residues while adhering to the principles of sustainability. Factors related to biodiversity conservation and soil acidification have been found to be limiting for logging residues and stumps biomass harvesting [10].

Assessing the energy potential of woody biomass in mountain forests is a relevant area of research around the world. According to the published research of mountain forests of Japan, industrial forests were selected, and the annual available volumes of wood biomass were assessed under conditions of favorable forest management. Industrial forests were considered in detail with the widespread introduction of geographic information systems. The annual volumes of forest woody biomass available for harvesting were estimated on the basis of the annual production potential. According to the research, the expected revenues exceeded all costs from forest planting to biomass harvesting within a 60-year rotation [11].

Exhibitory research carried out in the province of Biscay in northern Spain identified, quantified and characterized potentially available wood biomass of *Pinus radiata* for its use to meet local energy needs [12]. For this purpose, the information of the National Cadasters of Spain on quantification of the amount of sequestered CO_2 by forests on the basis of forestry statistics, was used.

A considerable attention is paid to the study of the energy potential of wood biomass in the United States. Severe droughts in recent years and catastrophic forest fires have led to the formation of significant stocks of dead wood in California's forests. This phenomenon justified the need to assess energy potential of dead wood biomass. The authors propose a new approach to the assessment of standing dead trees biomass based on aerial photography data and forest structure maps. Relationships between the intensity and duration of droughts and the accumulation of standing dead wood have been established, and its energy potential has been assessed [13].

The presented analysis of scientific publications allows to conclude that the use of woody biomass requires an environmentally balanced approach, a proper information support, and a well-defined target level of detail at management planning. Given the presence of the above preconditions, woody biomass of mountain forests is capable to secure local energy needs, contribute to national energy security, energy transition, as well as help fulfil the international climate-related treaties signed by Ukraine.

3. The purpose of this work

The purpose of this research is to assess biomass stock in Ukraine's forests as well as prospects for its utilization as an alternative source of energy while adhering to the basic foundations of sustainable development in forestry.

4. Materials and methods of research

The research covers forests of Ukraine, which cover an area of over 9.5 million hectares, where about 2 billion m^3 of trunk stock has been accumulated. According to the peculiarities of the climate, large landscape forms, soils and natural vegetation, the country is divided into five natural zones – Polissia, Forest-Steppe, Steppe, Carpathian and Crimean Mountains. The largest forest areas are concentrated in the northern and western parts of the country – in Polissia (mixed forests) and in the Ukrainian Carpathians. Coniferous stands cover 42 % of the total forested area, including Scots pine (*Pinus silvestris* L.), the share of which is 32 %, Norway spruce (*Picea abies* Karst.) and silver fir (*Abies alba* Mill.), which account for approximately 10 % of forested area. Hardwood forests account for 43 %, of which common oak (*Quercus robur* L.) and European beech (*Fagus sylvatica* L.) dominate over 32 % of the total forested area. Almost 15 % of the forested area is represented by softwood broadleaves and shrubs.

The age structure of forests is as follows: young stands – 32 %, mid aged – 44, maturing – 13, mature – 11 %. The mean age of stands is nearing 62 years. Artificially created (planted) stands make up more than 50 % of the total forested area. Forest productivity is quite high. The average stock per 1 ha of forested land reaches 220 m^3·ha^{-1}.

For research purposes, we have used experimental data collected on temporary sample plots according to the tested methodology and sampling technique [14]. The total amount of research data used is 3035 sample trees, 1743 sample sections of tree trunks and 809 samples of crown branches (Table 1).

Table 1. Quantitative description of the input research data.

Tree species	Quantity, pcs		
	sample trees	sample sections	
		trunk	crown branches
Scots pine	803	210	54
Pallasian pine	196	108	54
Norway spruce	279	150	99
Silver fir	200	243	99
Common oak	342	354	123
European beech	167	96	36
European hornbeam	37	27	8
European ash	64	124	62
Silver birch	198	42	79
European aspen	635	217	108
Black alder	114	172	87

Mathematical models of biomass expansion factors approbated in a number of scientific publications were used to establish quantitative values of biomass stock in Ukraine's forest [15, 16]. At the same time, to assess the energy potential of woody biomass as an alternative energy source, we used a dependence that reflects the sum of its structural components and has the following form:

$$PWB_{a,j} = PFW_{a,j} + PFR_{a,j} + PRWP_{a,j} \tag{1}$$

where $PWB_{a,j}$ – total energy potential of woody biomass in year a over the area j, m^3·annuum^{-1} or J·annuum^{-1}; $PFW_{a,j}$ – energy potential of fuel wood from stem wood in year a over the area j; $PFR_{a,j}$ –

energy potential of logging residues in year *a* over the area *j*; $PRWP_{a,j}$ – energy potential of wood processing residues in year *a* over the area *j*.

5. Research results

Using the information on biometric characteristics of forested areas obtained within the current State Forest Account of Ukraine, it was found that the total energy potential of biomass in the country is over 27 EJ (Table 2). The energy value of biomass was established on the basis of data by A. Shvidenko, S. Nilsson and M. Obersteiner [17].

More than 47 % of the energy stock in biomass of Ukraine's forests is concentrated in hardwood broadleaved stands, of which 26.8 % – in common oak stands. The share of the country's coniferous forests in the total biomass energy potential is 43.4 %, including 33.0 % of pine stands.

About 78 % of woody biomass stocks are concentrated in stem wood and tree canopy branches. These components of woody biomass are used in Ukraine as a renewable energy source. Other biomass components remain on forest plots for further biodegradation according to the common forestry practice applied in Ukraine. For this purpose, it is necessary to leave on sites a fraction of crown branches, that will promote restoration of forest soils fertility.

Table 2. Energy content in biomass of Ukraine's forests by components.

Tree species groups and dominant tree species	Energy content by biomass components, PJ					
	stem	branches	foliage	roots	understory	total
Conifers	8310.6	869.0	294.5	2029.4	261.5	11765.0
Including pine	6504.7	606.1	146.4	1448.3	241.8	8947.4
spruce	1534.1	219.9	123.9	477.4	14.8	2370.1
Hardwood broadleaves	8285.6	1940.0	164.1	2138.4	388.2	12916.4
Including oak	4997.5	924.4	104.6	999.5	248.4	7274.3
beech	2344.1	827.8	38.6	960.2	80.6	4251.3
Softwood broadleaves	1603.8	234.2	56.3	430.9	72.4	2397.7
Including birch	759.9	134.1	33.8	203.8	32.9	1164.5
aspen	75.1	7.2	3.2	30.4	6.6	122.4
Other tree species	23.2	5.4	1.6	7.2	4.9	42.3
Total	18223.3	3048.5	516.6	4605.9	727.1	27121.3

In Ukraine, more energy potential of woody biomass of forests is concentrated in mid-aged stands – 51.1 % (Table 3). In these stands commercial thinning is carried out, which allows to receive more than 1.6 million m^3 of wood annually, of which about 60 % is fuel wood. More than 18 % of woody biomass stocks are accumulated in mature and overmature stands, which is 4.9 EJ in energy equivalent. Final cuts are carried out in stands of this age group, allowing to source 7.5–8.5 million m^3 of merchantable wood, including more than 2 million m^3 of stem fuel wood. Sanitary cuts, which are aimed at improving sanitary condition of Ukraine's forests, are also an important component of forest management measures and activities. If necessary, they are prescribed in mid-aged and maturing stands. Taking into account the current sanitary condition of Ukraine's forests, about 10 million m^3 of wood is harvested annually in course of this kind of forest management activity, about half of this volume is fuel wood.

Systems of informational, legislative and technological measures in forestry and wood processing industry will facilitate sustainable growth of share of renewable energy from forest biomass in production of industrial products.

Based on the average annual volumes of merchantable wood harvesting and structural relationships between individual biomass components in stands, we have assessed the annual energy potential of woody biomass for Ukraine's forests.

Table 3. Energy content in biomass of Ukraine's forests by age groups.

Tree species groups and dominant tree species	Energy content in biomass by age groups, PJ					
	young	mid-aged	maturing	mature	overmature	total
Conifers	1608.0	5864.2	3035.1	1170.5	87.2	11765.0
Including pine	1195.0	4424.2	2562.2	738.8	27.1	8947.4
spruce	366.9	1223.8	414.6	321.3	43.6	2370.2
Hardwood broadleaves	971.5	7094.3	2008.6	1930.3	911.7	12916.3
Including oak	720.5	4735.1	934.9	703.8	180.1	7274.4
beech	148.7	2136.8	762.0	729.1	474.8	4251.3
Softwood broadleaves	98.6	893.8	605.6	606.8	192.9	2397.7
Including birch	59.7	472.3	357.2	261.1	14.2	1164.5
aspen	8.6	10.2	23.8	55.1	24.7	122.5
Other tree species	5.3	15.5	7.1	6.9	7.5	42.3
Total	2683.4	13867.8	5656.4	3714.4	1199.4	27121.3

This 93.9 PJ resource (Table 4) serves as a renewable energy source and is an important component of Ukraine's fuel and energy balance. More than 50 % of the potential is accounted for by stem fuel wood, and almost 25 % is logging residues biomass.

Table 4. Annual energy potential of woody biomass of Ukraine's forests.

Tree species groups and dominant tree species	Energy potential of woody biomass by structural components, PJ			
	stem wood	logging residues	wood processing residues	total
Conifers	30.20	13.76	11.03	54.99
Including pine	24.89	11.34	9.09	45.31
spruce	3.65	1.66	1.33	6.64
Hardwood broadleaves	16.23	7.40	7.00	30.63
Including oak	8.63	3.93	3.72	16.29
beech	3.51	1.60	1.51	6.62
Softwood broadleaves	4.34	1.98	1.87	8.19
Including birch	1.82	0.83	0.79	3.44
aspen	0.31	0.14	0.13	0.58
Other tree species	0.05	0.03	0.02	0.09
Total	50.82	23.16	19.92	93.90

Generally, almost 60 % of energy potential of woody biomass is concentrated in the north-western regions of Ukraine, another 30 % – in its central part and less than 10 % – in the south-eastern regions. At the same time, the forests of the Ukrainian Carpathians are characterized by the highest density of woody biomass per unit of forested area – more than 450 $KJ \cdot (m^2)^{-1}$.

6. Conclusions

Within the frame of this research we have assessed energy potential of woody biomass of Ukraine's forests. In units of primary energy its value is more than 27 EJ. At the same time, to facilitate securing energy needs of society, the annual potential of woody biomass, which can be used as a source of renewable energy, has been established. For this purpose, an annual potential of 93.9 PJ is available in Ukraine. More than 50 % of the woody biomass potential is accounted for by fuel wood from stem wood, and almost 25 % is biomass from logging residues. Almost 60 % of the potential is concentrated in the north-western regions of Ukraine, in particular in such forested areas as the Ukrainian Carpathians and Polissia. At present, to ensure the rational use of energy potential of woody biomass, regional target programs for replacement of fossil energy resources with wood biofuels are extremely important.

References

[1] Soshenskyi O, Zibtsev S, Gumeniuk V, Goldammer J G, Vasylyshyn R and Blyshchyk V 2021 The current landscape fire management in Ukraine and strategy for its improvement *Environmental & Socio-economic Studies* **9 (2)** pp 39–51

[2] Vasylyshyn R, Yurchuk Yu, Lakyda I and Bondarchuk R 2021 Energy content in woody biomass of Zhytomyr Polissia's forests *Ukrainian Journal of Forest and Wood Science* **12 (1)** pp 25–33 doi: 10.31548/forest2021.01.003

[3] Vasylyshyn R 2018 *Environmental and energy potential of forests in Ukrainian Carpathians and its sustainable use* (Kyiv: LLC Komprint) p 305

[4] Lakyda P, Sytnyk S, Lovynska V, Lakyda I and Gritsan Y 2020 Crown biomass of Scots pine and black locust in Northern Steppe of Ukraine *Forestry Ideas* **26(2)** pp 471–484

[5] Sytnyk S, Lovynska V and Lakyda I 2017 Foliage biomass qualitative indices of selected forest forming tree species in Ukrainian steppe *Folia Oecologica* **44(1)** pp 38–45

[6] Lovynska V, Lakyda P, Sytnyk S, Lakyda I, Grithan Y and Hetmanchuk A 2019 Stem production of Scots pine and black locust stands in Ukraine's Northern Steppe *Journal of Forest Science* **65(12)** pp 461–471 doi.org/10.17221/92/2019-JFS

[7] Zverkovskyy V, Sytnyk S, Lovynska V, Kharytonov M, Lakyda I, Mykolenko S, Margui E and Gispert M 2018 Remediation potential of forest forming tree species within northern steppe reclamation stands *Ekologia Bratislava* **37(1)** pp 69–81 doi.org/10.2478/eko-2018-0007

[8] Gonzalez-Salazar M, Venturini M, Finkenrath M, Poganietz W-R, Morini M, Pinelli M and Ruggero P 2013 Methodology for Biomass Energy Potential Estimation: Projections of Future Potential in Colombia *Renewable Energy* **69** pp 488–505

[9] López-Rodríguez F, Sanz-Calcedo J and Moral García F 2019 Spatial Analysis of Residual Biomass and Location of Future Storage Centers in the Southwest of Europe *Energies* **12(10)** doi.org/10.3390/en121019785

[10] de Jong J, Akselsson C, Egnell G, Löfgren S and Olsson B 2017 Realizing the energy potential of forest biomass in Sweden – How much is environmentally sustainable? *Forest Ecology and Management* **383** pp 3–16 doi.org/10.1016/j.foreco.2016.06.028

[11] Aruga K, Murakami A, Nakahata C, Yamaguchi R, Saito M and Yoshioka T 2014 Estimating Annual Available Amounts of Forest Biomass Resources with Total Revenues and Costs during the 60-Year Rotation in a Mountainous Region in Japan *Croatian Journal of Forest Engineering* **35** pp 125–139

[12] Mateos E, Garrido F and Ormaetxea L 2016 Assessment of Biomass Energy Potential and Forest Carbon Stocks in Biscay (Spain) *Forests* **7(4)** doi.org/10.3390/f7040075

[13] Tubbesing C L, Lara J D, Battles J J, Tittmann P W and Kammen D M 2020 Characterization of the woody biomass feedstock potential resulting from California's drought *Scientific Reports* **10** doi.org/10.1038/s41598-020-57904-z

[14] Lakyda P 2002 *Live biomass of Ukraine's forests* (Ternopil: Zbruch) p 256

[15] Shvidenko A, Lakyda P, Schepaschenko D, Vasylyshyn R and Marchuk Yu 2014 *Carbon, climate and land-use in Ukraine: forest sector* (Korsun-Shevchenkivsky: Gavryshenko) p 283

[16] Shvidenko A, Schepaschenko D, Nilsson S and Bouloui Y 2008 *Yield tables, growth and productivity models of stands of the main forest forming tree species of Northern Eurasia* (Moscow: Moscow printing house № 6) p 887

[17] Shvidenko A, Nilsson S and Obersteiner M 2004 Wood for bioenergy in Russia: Potential and Reality *Wood Energy in the Industrialized World* **1** p 323–340

IOP Conf. Series: Earth and Environmental Science 1042 (2022) 012011 doi:10.1088/1755-1315/1042/1/012011

The performance analysis of modified generic model for Lithium-ion batteries

Shubha Rao K*, **Karanam Vasudha**[2]

BNM Institute of Technology, Bengaluru, India

*Email: shubharaok@bnmit.in

Abstract. Battery technology has been widely used in various applications such as electric vehicles, smart grids, renewable energy systems, and sustainable power applications in recent years. The type, rating, and operating characteristics of batteries vary from one application to another. For example, Electric Vehicles (EVs) need lightweight batteries capable of producing fast energy and withstanding frequent charge/discharge cycles. Renewable energy applications require batteries that can store large energy in the time of excessive electric power generation and can substitute renewable energy sources to supply power in times of non-generation. As batteries are an electro-chemical phenomenon and are generally expensive, it is essential to establish an effective battery model to analyze its behavior under different operating conditions. In this regard, this paper presents a comprehensive study of various battery modeling approaches widely used for the characterization of batteries employed as energy sources for electric vehicles. A simple modified generic battery model is developed and simulated for the ANR26650M1 Li-ion battery in a MATLAB-Simulink environment. The modified generic model includes a polarization voltage term to represent the open-circuit voltage accurately and the polarization resistance term is slightly modified. The model parameters are deduced from the data sheet provided by the battery manufacturers. The discharge curve obtained through simulation is validated with the one given in the datasheet.

1. Introduction

In recent years people all over the world are preferring to have private vehicles as they provide personal control, comfort, status, ease for the movement of older people and due to lack or poor quality of public transport. Since Internal Combustion Engine Vehicles (ICEV) are widely used, this resulted in high demand for fossil fuels. The use of fossil fuels is increasing environmental pollution, on the other side, the lack of resources is increasing the cost of petrol and diesel. To control the pollution, several new policies and rules were launched to develop green cars, this made researchers develop cars that run on electricity called electric vehicles (EV). The battery is a very important component in any electric vehicle, as it acts as an energy source. EV range depends on the battery, therefore several batteries are to be connected in series and parallel. The proper balancing of these batteries is required to utilize the battery bank completely. Cell balancing and precise estimation of the State of Charge (SOC) are the key parameters to achieve this.

A Battery model is nothing but imitating the behavior of the battery. A Battery exhibits nonlinear effects during charge/discharge; this increases the complexity of modeling the battery behavior. An accurate battery model should consider the effect of temperature, SOC, State of Health (SOH), and current in its behavior and accurately reproduce its nonlinear phenomenon. Battery modeling is widely used to understand the behavior of the battery for different load cycles, analyze the cell balancing in

EVs, and analyze the overall performance of the electric vehicle. This way, the system designers can optimize the performance of the overall system.

A simple basic battery model that can be developed is the shepherd model. However, the basic model is applicable for steady-state current only and has the drawback of algebraic loop problems during simulation. This paper proposes a modified model which overcomes the drawbacks by including a polarization voltage term $* \frac{Q}{it+0.1*Q} * it$, in the model equation to remove algebraic problems during simulation.

In this paper, an overview of various battery modeling methods is discussed, in particular, the electric circuit model and parameterization techniques are presented in detail. Section-2 provides a brief overview of various battery modeling approaches. The electric circuit models are discussed in detail in section-3. Different ways of determining battery modeling parameters through experimentation are discussed in section-4. The modeling procedure and simulation results are discussed in section-5. The conclusions are drawn in Section-6.

2. Battery Modelling Methods

There has been tremendous research work being carried out in the field of modeling batteries for different areas of applications. The battery models are broadly categorized into i) electrochemical model, ii) mathematical model, and iii) electrical equivalent circuit model. The type of modeling employed depends on the area of application. For instance, electrochemical models are widely used by researchers and manufacturers during the development phase. Mathematical models are widely used for research purposes and it's challenging to derive mathematical models for batteries as they are a complex electrochemical phenomenon. Electrical equivalent models are simple, easy to derive, and less accurate. However, equivalent circuit models are more than sufficient for performance analysis and prediction of battery behavior in electric systems.

2.1. Electrochemical model

The electrochemical models are derived by directly considering the electro-chemical reactions happening inside the battery. The parameters of the electrochemical model are plentiful and hence it is time-consuming to determine all parameters. Hence only a few parameters are obtained through experimentation [1]. These models comprise a large number of partial differential equations (PDEs) which leads to the requirement of enormous memory space and computational effort. So it may be difficult to perform a real-time simulation with such a complicated electrochemical model to validate EV-BMS. This problem has been successfully dealt with by many researchers by reducing the complexity of models that can analyze battery performance and estimate SOC with an acceptable degree of fidelity [2].

2.2. Mathematical and analytical models

A mathematical model is a simplified electrochemical model where battery terminal voltage is represented in terms of SoC and battery current. There are several mathematical models like Shepherd's equation, Unnewehr universal model, Nernst model, and combined model. In Peukert's modified equation, the battery model is improved by including integrating current, it is based on empirical data/experimental data, and is also called an analytical method.

2.3. Equivalent Electrical circuit Model

The electrical circuit representation of batteries consists of Voltage/Current Sources, Passive elements such as Resistances, Capacitances, and sometimes inductances connected in series and/or parallel. It can effectively estimate battery parameters such as SOC, SOH, Life cycle of the batteries by sensing cell currents and voltages. These models are accurate enough to be used with other sub-systems to perform co-simulation and analyse the performance whole systems. It is the most preferred model to perform simulation of complete electric vehicles as it can be easily integrated. However, the basic electrical circuit models do not take into account non-linearity effects in battery behavior, resulting in

IOP Conf. Series: Earth and Environmental Science 1042 (2022) 012011 doi:10.1088/1755-1315/1042/1/012011

the inaccurate prediction of battery performance [3-4]. Many advanced electric circuit models are presented in [5-10] which are simple but can accurately predict the battery's remaining capacity. The basic classification of electrical circuit modeling is explained in the next section.

3. Electric Circuit Modelling

Electrical circuit models (ECM) are divided into three types: i) Rint model, ii) Thevenin model, iii) Double polarization (DP) model, and iv) Runtime model.

3.1. Rint Model

Figure 1 shows the Rint model which is the simplest type of model that consists of an open-circuit voltage $V_{OC,}$ in series with its internal resistance, R_i which represents the polarization of the battery. The open-circuit voltage V_{OC} is measured by open-circuit of load terminals whereas the R_i value is obtained from measurements with connected load, for a fully charged battery under consideration. However, the Rint model has the drawback of assuming the internal resistance of the battery as constant and it does not consider the variation of internal resistance with SOC. It provides information on the DC characteristics of the battery but does not provide transient characteristics. Hence it is difficult to capture the charge transfer phenomenon of the battery in the Rint model.

3.2. Thevenin Model

Figure 2 shows the Thevenin battery model which is a modification of the Rint Model with the addition of one or more parallel RC circuits. The Parallel RC network represents the dynamic behavior of the battery. The non-linear transient resistance R_T represents the battery's contact resistance whereas the non-linear transient capacitance C_T represents the capacitance of the parallel plate. This type of ECM can be parameterized using a hybrid pulse power characterization (HPPC) test schedule.

3.3. DP Model

DP model shown in figure 3 consists of an open-circuit voltage (V_{OC}), an internal resistance (Ri), and two parallel RC circuits. With only two parallel RC circuits, it can model the dynamics of lithium batteries quite accurately. Moreover, the DP model is the best model among the models that were widely used. It has low errors and low RMSE (Root Mean Square Error) among other models with the lowest complexity of the models. V_{OC} is the open-circuit voltage which is usually a function of the state of charge (SOC) of the battery, Ri is the internal resistance of the battery, R_{t1} and C_{t1} is the parallel RC circuit that describes the fast dynamic response of the battery, finally, R_{t2} and C_{t2} are the parallel RC circuit that describes the slow dynamic response of the battery.

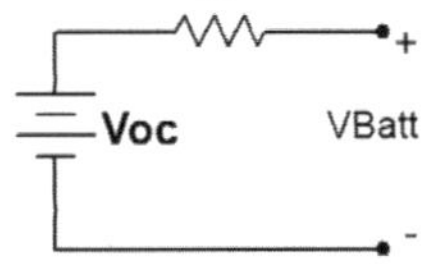

Figure 1. Rint Model

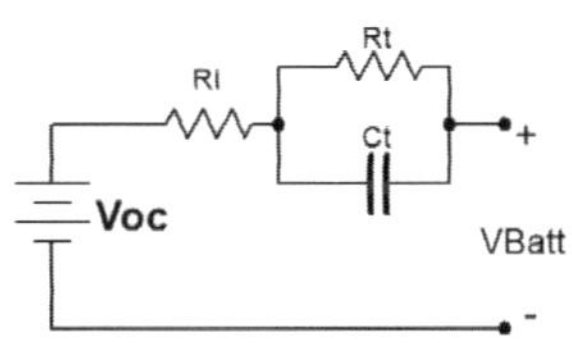

Figure 2. Thevenin Model

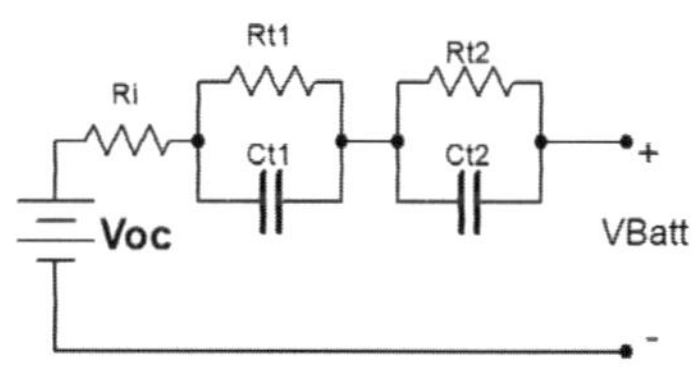

Figure 3. DP Model

Electrical models are easy to develop and simple as they comprise a voltage source, and passive components such as resistors, capacitors, and probably inductors. The Electrical models are widely used to study voltage/current characteristics, state of charge, and temperature effects in a battery as the accuracy obtained are sufficient for many applications including EV systems [11-14]. But a major drawback of electrical modeling of battery is that the parameter identification process is tedious and require sophisticated testing equipment. The parameterization of the battery is dealt with in brief in the next section.

IOP Conf. Series: Earth and Environmental Science 1042 (2022) 012011 doi:10.1088/1755-1315/1042/1/012011

4. Parameterization of battery

The parameters of the battery model can be determined accurately by experimentation, or from the data given in the datasheet. The popular way to get the parameters is by experimentation using the current pulse technique also called Hybrid Pulse Power Characterization Test (HPPC). During this test initially, the battery is completely charged, then made to discharge partially, and then it is allowed to rest for a while after this proper open-circuit voltage (Voc) is measured, and other required information is measured. This procedure is repeated until the battery discharges completely. This test data is used to determine many parameters like peak power, available energy, voltage response curves, battery modeling parameters, etc. This test is performed during many research works and the works published in papers are presented here:

In paper [15], this test is performed on a LiFePO4 lithium battery by using DP circuit modeling and is performed in Scilab/Xcos, where polynomial equations are used to represent the parameters of the circuit. The polynomial equations are obtained by curve fitting the experimental data which is obtained from the HPPC test. Here the battery is connected to a sensor cable and power cable, and battery analyzer BTS8 10A, 30V records voltage and current at 25°C while charging and discharging the battery. Test data is used to develop the model and it shows satisfactory results when compared with experimental data.

In paper [16] the same test is done to obtain the relation between the open-circuit voltage and SOC. From SOC, the battery model parameters shown in fig.2 are determined. Here, all the parameters are obtained as polynomial expressions and are SOC's functions. It is also said that the model obtained is applicable for different lithium batteries from the same manufacturer under the same experimental conditions.

In this paper, A123 Li-iron-phosphate ANR26650M1 is modeled by estimating the parameters required for modeling from the discharge curves and data provided in the datasheet.

5. Simulation Results and Discussions

The simulation is performed using a generic dynamic model that is a combination of a modified shepherd equation and the Rint model. In this model, the shepherd equation is used to calculate the open-circuit voltage required by the Rint model. The modified shepherd formula uses only battery SoC as a state variable to represent battery voltage variations and it incorporates polarization voltage to avoid algebraic loop problems during Simulink simulation.

Equation (1) and (2) represents the battery open-circuit voltage during charge and discharge respectively. The parameters in equations (1) and (2) are determined from the data sheet.

$$\text{Discharge: } E_{batt} = E_0 - i \times R - \left(K \times \frac{Q}{Q-it} \times it \right) - \left(K \times \frac{Q}{Q-it} \times i^* \right) + \left(A \times e^{-(B \times it)} \right) \quad (1)$$

$$\text{Charge: } E_{batt} = E_0 - i \times R - \left(K \times \frac{Q}{it+0.1Q} \times it \right) - \left(K \times \frac{Q}{Q-it} \times i^* \right) + \left(A \times e^{-(B \times it)} \right) \quad (2)$$

Where

E_{batt}: Battery Voltage (V)
$A=V_{full} - V_{exp}$ (V): drop in voltage in the exponential area
$B=3/Q_{exp}$ (Ah^{-1}) : Charge at the end of the exponential area
$K= (V_{full}-E_{nom}+A*(e^{-(B*Q_{nom})}-1)*(Q-Q_{nom})*Q_{nom}$ (V/ Ah^{-1}) : polarization constant
$Eo=V_{full} + K + R.i-A$ (v) : Fully charged voltage
Q: Capacity of the battery in Ah
it: Actual charge of the battery in Ah
R: internal resistance (Ω)
i = battery current in A

i^* = filtered current (A)

$K \times \frac{Q}{it + 0.1Q} \times it$:Polarization Voltage

$K \times \frac{Q}{Q - it}$: Polarization Resistance

Figure 4 shows the nonlinear battery model used to simulate the characteristics of the A123 Li-iron-phosphate ANR26650M1 battery. It consists of a controlled voltage source in series with constant internal resistance. The model parameters are determined by using the discharge curve given in the datasheet of the battery. Figure 5 shows a typical discharge curve of a battery which starts with an exponential area where battery voltage reduces exponentially from a full charged value. The nominal area in the discharge curve represents an operating region of the battery where battery voltage remains at a nominal value while being discharged. The terminal voltage falls sharply after the nominal zone which needs to be avoided. [17-19].

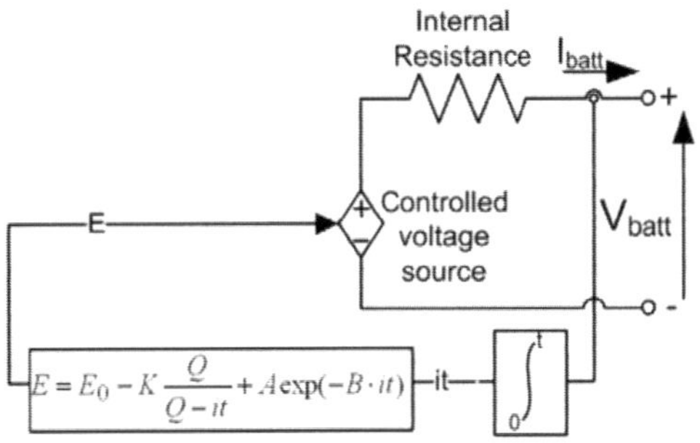

Figure 4. Non-linear Battery Model

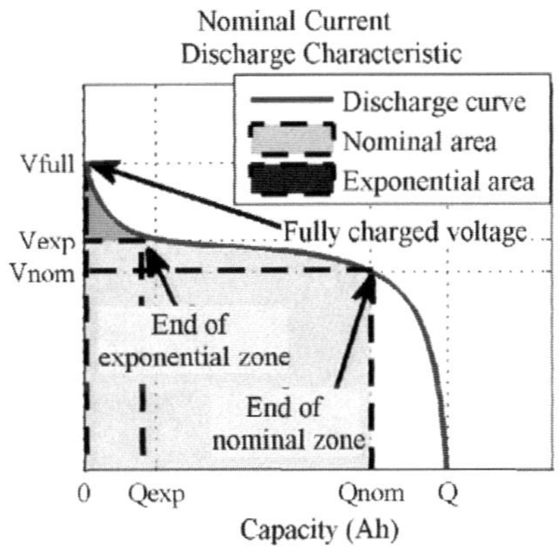

Figure 5. General discharge curve

The values obtained for the considered A123 Li-iron-phosphate ANR26650M1 battery are K=0.02, A=0.03 V, E_0=3.443 V, B=13.04 Ah^{-1}, Q=2.3 Ah, and R=0.01 Ω. Figure 6 shows the battery model developed in MATLAB/Simulink. Figure 7 shows the simulated battery voltage v/s SOC characteristics at 25°C for a nominal current of 2.3A. The V_{nom} and Q_{nom} obtained in simulation are 3V and 88% respectively whereas in the datasheet it is 3.1V and 90%. This validates the battery model developed and its accuracy. The simulation is done with a discharge current of 2.3A at 25° C. Figure 9 shows the Discharge characteristics obtained by simulating the model at three different battery currents of 1A, 10A, and 30A. The battery voltage decreases sharply as the discharge current increases. Figure 10 shows the SOC, battery current, and terminal voltage obtained through simulation at a discharge current of 2.3A. It is observed that the safe region of operation for a battery is 20% to 85% of SOC.

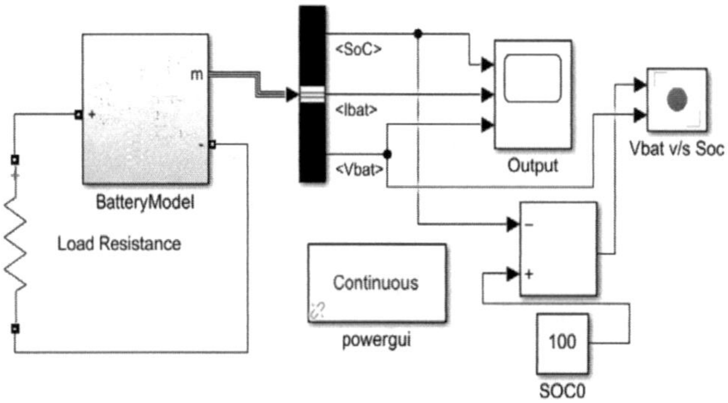

Figure 6. Battery Modeling using Simulink tool

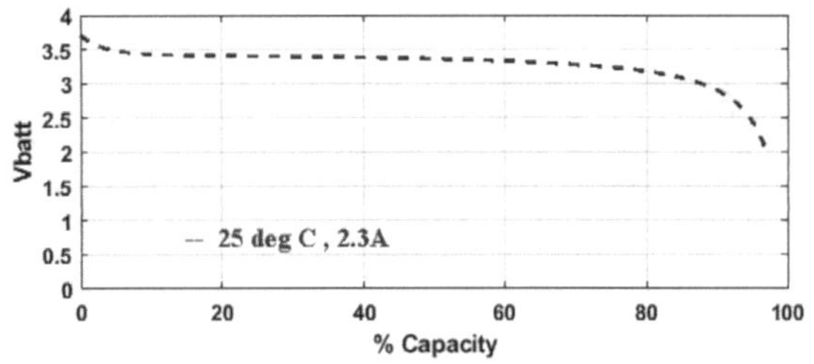

Figure 7. The discharge curve of the battery through simulation for 25 dag C

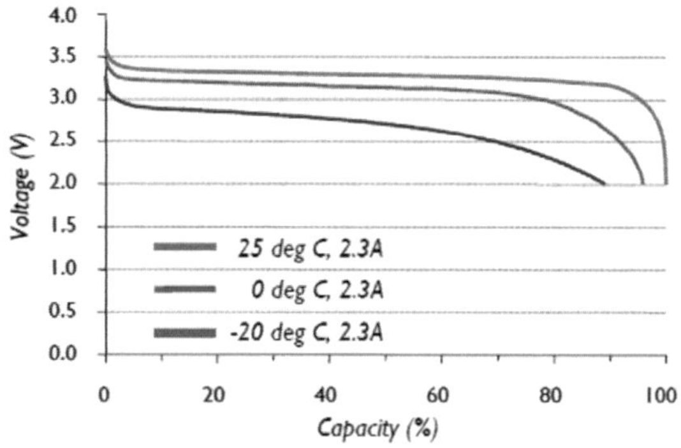

Figure 8. Discharge curves from the data sheet

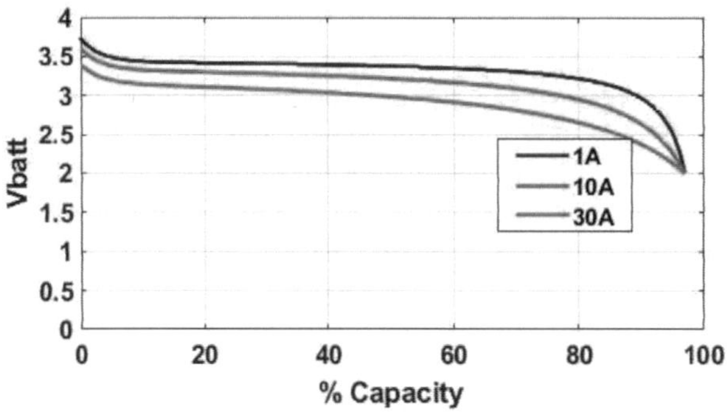

Figure 9. Battery Discharge Characteristics at different Discharge currents (1A, 10A, and 20A)

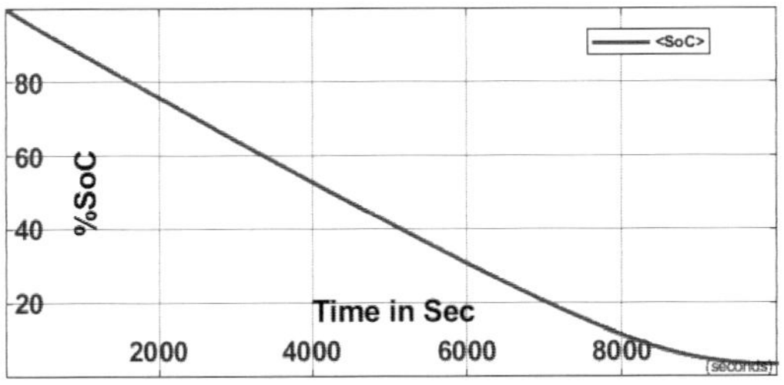

(a)

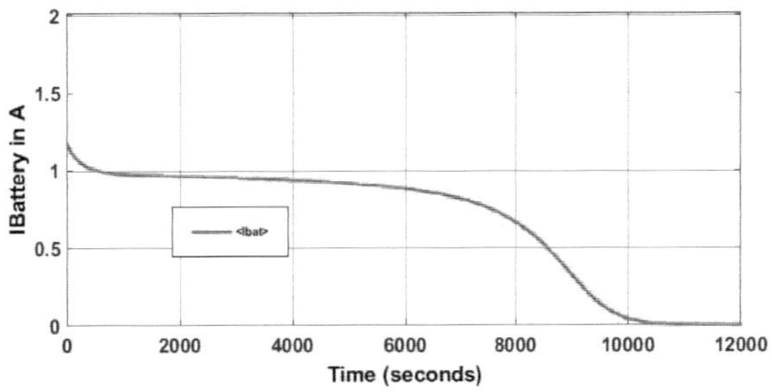

(b)

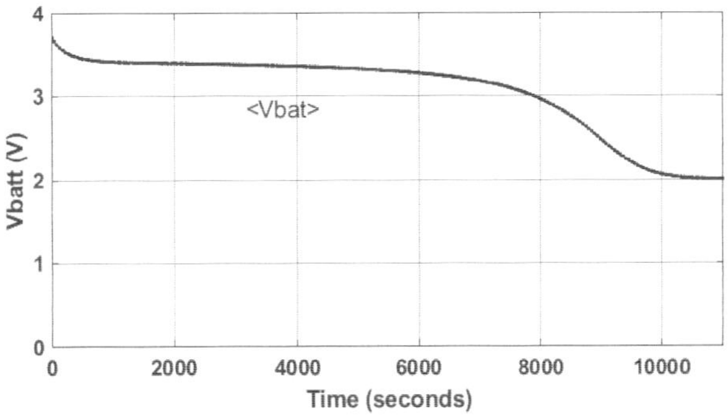

(c)

Figure 10. Battery (a) SOC (b) Current (c) Voltage during discharge

6. Conclusion

This paper presented a brief overview of the modeling of batteries with a focus on electrical circuit modeling. It illustrated how electrical circuit models can reflect the typical electrochemical characteristics of batteries. The advantages and disadvantages of each electrical circuit model were outlined. Further, the parameterization and implementation technique used for these models is discussed. At last, a generic battery model is simulated for the Li-iron-phosphate ANR26650M1 battery using a generic dynamic model approach. The variation of open-circuit voltage with respect to % capacity is obtained for a nominal current of 2.3A and the results are found to be satisfactory with the data provided in the datasheet. The graph of battery voltage for various values of the discharge current is also obtained and it is observed that the battery operates satisfactorily in the range of 20% to 80% of battery capacity.

References

[1] Jinhao Meng, Guangzhao Luo Mattia Ricco, Maciej Swierczynski, Daniel-Ioan Stroe and Remus Teodorescu, 2018 Overview of Lithium-Ion Battery Modeling Methods for State-of-Charge Estimation in Electrical Vehicles Journal of Applied Sciences 8 Issue 5 pp 1-17

[2] R. Ahmed, M. E. Sayed, I. Arasaratnam, J. Tjong and S. Habibi, Sept. 2004 Reduced-Order Electrochemical Model Parameters Identification and SOC Estimation for Healthy and Aged Li-Ion Batteries Part I: Parameterization Model Development for Healthy Batteries in IEEE Journal of Emerging and Selected Topics in Power Electronics 2 Issue 3 pp. 659 – 677.

[3] Yin Jin, Wenchun Zhao, Zhixin Li, Baolong Liu and Liu Liu 2021 Modeling and Simulation of

Lithium-ion Battery Considering the Effect of Charge-Discharge State *Journal of Physics: Conf. Ser.* **1907** (IOP Science) pp 1-7

[4] Felix Heinrich1, Patrick Klapper and Marco Pruckner 2021 A comprehensive study on battery electric modeling approaches based on machine learning *Proc. of the 10th DACH+ Conference on Energy Informatics* 4 Supplement 3 pp 1-17

[5] Yongxue Ye, Hongyan Ma, Jingjian Yang 2020 Research on Accurate Model of Lithium Battery *IEEE Chinese Control And Decision Conference (CCDC)* pp 1966-1970

[6] G.Marin-Garcia, G.Vazquez-Guzman, J.M.Sosa, Adolfo R. Lopez, P.R. Martinez-Rodriguez, D. Langarica 2020 Battery Types and Electrical Models: A Review *IEEE International Autumn Meeting on Power, Electronics and Computing* (Ixtapa, Mexico) pp 1-6

[7] Srinivas Singirikonda and Y P Obulesu 2020 Battery modelling and state of charge estimation

[8] methods for Energy Management in Electric Vehicle-A review IOP Conf. Series: Materials Science and Engineering **937** pp 1-13

[9] Mircea Ruba, Raul Nemeş, Sorina Ciornei, and Claudia Marţiş 2020 Parameter Identification, Modeling and Testing of Li-Ion Batteries Used in Electric Vehicles *Applied Electromechanical Devices and Machines for Electric Mobility Solutions* (IntechOpen Book Series) pp 1-19

[10] P Chatterjee, J Singh, R Singh, Y A R Avadh and S Kanchan 2020 Electric Vehicle Modeling in MATLAB and Simulink with SoC & SoE Estimation of a Lithium-ion Battery *IOP Conf. Series: Materials Science and Engineering* (Mathura India) **1116** pp 1- 18

[11] V. Castiglia, R. Miceli, G. Ala, P. Romano, F. Viola, G. Giglia, A. Imburgia, G. Schettino 2019 Modelling, simulation and characterization of Li-Ion battery cell *IEEE 5th International Forum on Research and Technology for Society and Industry* (Florence, Italy) pp 1-6

[12] Gaizka Saldaña, José Ignacio SanMartín, Inmaculada Zamora, Francisco Javier Asensio, and Oier Oñederra 2019 Analysis of the Current Electric Battery Models for Electric Vehicle Simulation *Energies* **12** issue 14 pp 1-27

[13] Hartmut Hinz 2019 Comparison of Lithium-Ion Battery Models for Simulating Storage Systems in Distributed Power Generation *special issue of Inventions "Inventions and innovation in Energy and Thermal/Fluidic Science"* **4** Issue 3 pp 1-22

[14] Meilisa Dewi Kharisma, Muhammad Ridwan, Arinata Fatchun Ilmiawan, Ferdaus Ario Nurman, Saiful Rizal 2019 Modeling and Simulation of Lithium-Ion Battery Pack Using Modified Battery Cell Model *6th International Conference on Electric Vehicular Technology* (Bali, Indonesia) pp 25-30

[15] Mohd Khair Hassan, Mhd kheir almasri, Azura Che Soh,Nashiren Farzilah Mailah,Mohammad Hamiruce Marhaban 2019 Modelling of A18560 Lithium-ion Battery Based on System Identification Method *IEEE Conference on Sustainable Utilization and Development in Engineering and Technologies* (Penang, Malaysia) pp 205-208

[16] M. Chen and G. A. Rincon-Mora, 2006 Accurate electrical battery model capable of predicting runtime and I-V performance *IEEE Trans.Energy Conservation* **21**, no. 2 pp. 504–511.

[17] Marian Tomasov, Martina Kajanova, Peter Bracinik, David Motyka 2019 Overview of Battery Models for Sustainable Power and Transport Applications *Transportation Research Procedia* **40** (Slovak Republic: Elsevier) pp 548–555

[18] Sigit Arianto, Rietje Y. Yunaningsih, Edi Tri Astuti, and Samsul Hafiz 2006 Development of Single Cell Lithium-Ion Battery Model Using Scilab/Xcos *Conference Paper in AIP Conference Proceeding* pp.1-6.

[19] Olivier Tremblay, Louis-A. Dessaint, and Abdel-Illah Dekkiche 2007 A Generic Battery Model for the Dynamic Simulation of Hybrid Electric *Vehicles IEEE Vehicle Power and Propulsion Conference* pp 284-289.

[20] Jinhao Meng, Guangzhao Luo, Mattia Ricco, Maciej Swierczynski, Daniel-Ioan Stroe and Remus Teodorescu 2018 Overview of Lithium-Ion Battery Modeling Methods for State-of-Charge estimation in Electrical Vehicles *Journal of Applied Sciences* **8** Issue 5 pp 1-17

IOP Conf. Series: Earth and Environmental Science 1042 (2022) 012012 doi:10.1088/1755-1315/1042/1/012012

An overview of Lithium-Ion batteries for electric mobility and energy storage applications

Ganesh Sankaran*, **S. Venkatesan**

Department of Mechanical Engineering, Vinayaka Mission's Kirupananda Variyar Engineering College, Salem, Tamilnadu, India

*Email: sganesh6@gmail.com

Abstract: The battery is the key source of green energy for vehicle movement or powering residential / industrial buildings. The increase in energy demand requires larger battery capacity and energy density to meet power requirements in mobility and stationary energy storage applications such as in emergency power backup, solar power storage, portable power packs, etc., In recent years, the technology used in batteries has improved beyond expectations, mainly due to various cell chemistry. It is essential to understand the basic design and cost challenges in battery design, selection , cell chemistry advancements and on-going research areas in battery cell development. The study uses a detailed literature review, technical comparison of various cell chemistries & applications and identification of key design parameters using technical data comparison. The study reveals that Lithium batteries have an advantage over other cell chemistries due to its specific energy density, cost, scale of production in mobility and energy storage applications.

1. Introduction

The battery market is growing with many cell manufacturers investing in Giga factories to meet the Mobility & Power sector demand. The cost of the battery ($/kWh), on the other side, is declining rapidly due to the demand pattern. Europe has invested heavily in research, manufacturing active materials, applications, and recycling. China is leading EV battery pack manufacturing, followed by Europe. In terms of an Electric Vehicle (EV) mobility space application, it is dominated by China (77%), United States of America (9%) and Europe (6%) as mentioned in [1]. The use of batteries in second-life applications is growing due to environmental, social, and economic pressures on safe handling and disposal. At this stage, it is essential to get an overview on growing battery cell technology, applications. This is due to the fact that energy produced from fossil fuels posing a huge challenge like stringent emission norms and sharp increase in cost & operating comfort. This paper will discuss two major application areas.

2. Mobility Applications

In a Battery Electric Vehicle (BEV), the battery is the primary source of energy for vehicle movement. It can be of various sizes and capacities based on the requirement and range. In an EV, the material cost of the storage battery is about 40% to 50% of its full vehicle material cost; thus, the affordability of EVs is directly proportional to a battery's affordability. There are different types of cell chemistries available in the market [2], but Lithium Iron Phosphate (LFP) and Nickel Manganese Cobalt (NMC) play a significant role in India in the EV field. This is because they are high energy-specific storage in less space and are highly efficient. Moreover, NMC is more reliable than LFP, and in energy storage, all over the globe, NMC dominates cell chemistry [2].

Lithium-ion batteries have been considered widely in the application space [3] due to the technological growth in battery chemistry. Li-ion's choice is evident due to improved safety, stability, cost, and higher energy density, the preferred design parameters for battery design [4]. The present Li-Ion can provide 200-300 Wh/kg of energy density. In addition, the increase in energy demand of BEV requires a larger battery capacity and energy density to meet power and range requirements. The Table.1. shows the relationship between energy density by Volume and Weight.

Table.1. Comparison of Lithium Chemistry Batteries [3]

Chemistry	Typical Voltage	Specific Energy	Cycle Life	Charge (C rate)	Discharge (C rate)	Durability in High Temp. Environments	Thermal Runaway Temp.	Production Costs
Lithium Cobalt Oxide (LiCoO22)	3.7 v	150 - 200 Wh/kg	500 - 1000	0.7 – 1.0 C Charges to 4.2 v	1C,2.5 v Cut-off	Low	150°C (302°F)	Low
Lithium Manganese Oxide (LiMn2O4)	3.7 v	100 - 150 Wh/kg	300 -700	0.7 – 1.0 C Typical, 3C max; Charges to 4.2 v	1C typical; 10C on some cells ,2.5 v Cut-off	Low	250°C (480°F)	Low
Lithium Nickel Manganese Cobalt Oxide (LiNiMnCoO2)	3.6 v	150 - 200 Wh/kg	1000 - 2000	0.7 – 1.0 C Charges to 4.2 v	1C typical; 2C on some cells ,2.5 v Cut-off	Low	210°C (410°F)	Low
Lithium Nickel Cobalt Aluminum Oxide (LiNiCoAlO2)	3.6 v	155 - 260 Wh/kg	500	0.7 C Charges to 4.2 v	1C typical;, 3.0 v Cut-off	Low	150°C (302°F)	Medium
Lithium Iron Phosphate (LiFePO4)	3.2 v	90 -125 Wh/kg	1000 - 3000	0.5 C – 4C Charges to 3.7 v	1C typical; 30C on some cells ,2.5 v Cut-off	High	270°C (518°F)	Low

2.1. Selection of Right Parameters for the Battery Pack

The cell chemistry determines the performance characteristics of the Li-ion battery pack. Due to recent advancements and research, many cell chemistries have been introduced considering various application demand. The most common cell chemistries are lithium cobalt oxide (LCO), lithium nickel cobalt aluminum oxide (NCA), lithium nickel manganese cobalt oxide (NMC), lithium iron phosphate (LFP), lithium titanate oxide (LTO), lithium manganese oxide (LMO) and others like NiMH, Flow batteries [5] are available for choice.

After the choice of the right chemistry and suitable application, it is necessary to ensure the cell specification, as shown in Table.2. Besides, the trade-offs among Lithium-ion technologies are essential to understand from the designer's point of view [6] as per Figure.1.

Table.2. Battery Pack Design - Selection Parameters

Size of the battery in length, width, height (mm)	Battery Capacity in kWh
Charging & Discharging Current Rating	Battery Chemistry
Operating Voltage and Voltage Range	Battery Weight (kg) & Battery Volume (l)
Operating Temperature	Pack specific power & DoD
Pack Energy Density (Wh/l)	Pack Specific Energy (Wh/kg)
Cell Type & Cell Capacity (Ah)	Thermal Management
IP Class	Electrical Architecture

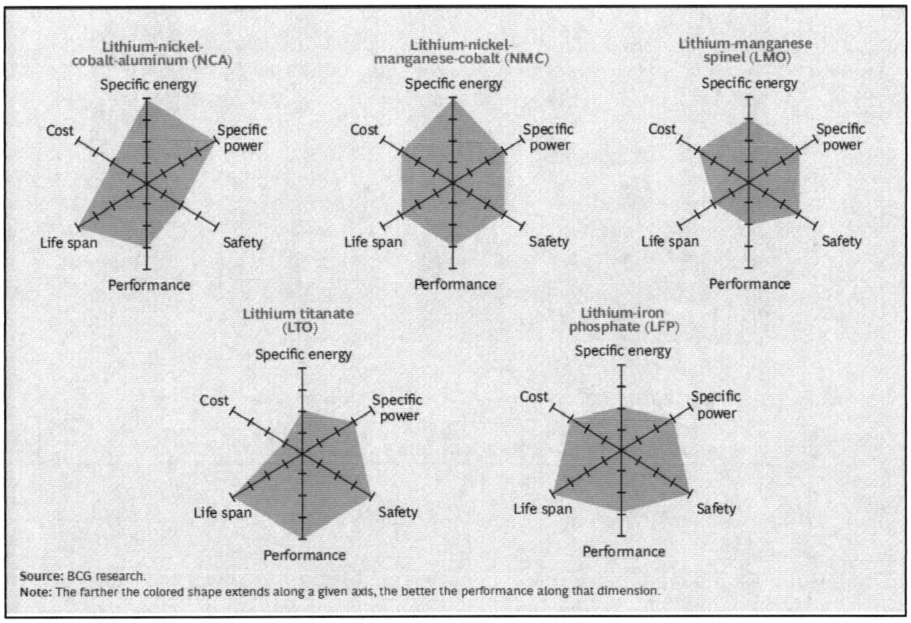

Figure 1. Trade-offs among the five Li-ion Battery Technologies

To understand the battery efficiency, it is essential to know the critical battery performance indicators [7] definition and validation using standard test procedures as shown in Table.3. The study shows how important it is to choose the suitable material for battery chemistry to achieve the proper energy density, and depicted in Figure.2. Higher-nickel fatty NMC cells, providing more energy density (vehicle range) have been preferred by EVs with longer ranges and addressing consumers' range anxiety. Therefore, the Shift is happening from NMC 532 to NMC 622 to NMC 811, as highlighted in [8].

Table.3. Battery performance indicators

Parameter	Metric	Description
Investment / Capital	INR/kWh	The investment required for battery procurement
Safety		Protection against thermal issues
Cycle Life	Number of Cycles	Number of times battery can be charged from 100% to 20% until capacity fades to 80% of its installed capacity
Volumetric/ Gravimetric Energy Density	Wh/L & Wh/kg	The amount of energy stored per liter and per kg (volume & weight)
Power Density	C - rate	It is a measure of power output per unit volume
Charging Time	C - rate	It is the rate at which a battery is charged relative to its maximum capacity
Reliability		The capability to operate in different temperature profiles (low/medium/high)
Others		Warranty & Maintenance costs, Calendar life, Charging efficiency

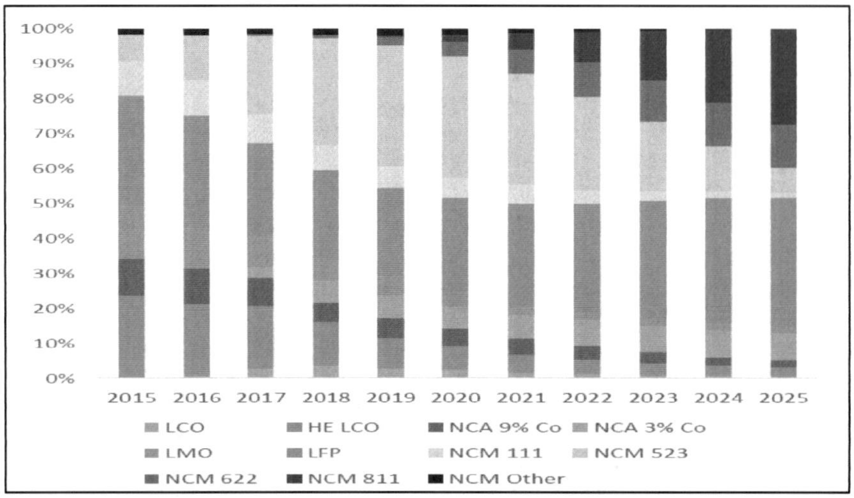

Figure 2. Cathode usage by type, (% market share) 2015-2025 [8]

The inference from the above chart is the reduction in LMO usage and LFP usage due to reduced energy density availability and shorter cycles used only in the first generation EVs. The increase in Nickel content in the chemistry NMC helps achieve higher energy density, thereby addressing good vehicle range and customer range anxiety issues as per [8]. The challenge in moving to higher Ni content cathode is the demand for specific raw materials [9], manufacturing processes & quality management systems [8]. Figure.3. illustrates the design factors to be considered by any designer or OEM before finalizing the form factor or geometry of the battery pack. These factors have a direct influence on integration into vehicle platforms or floors.

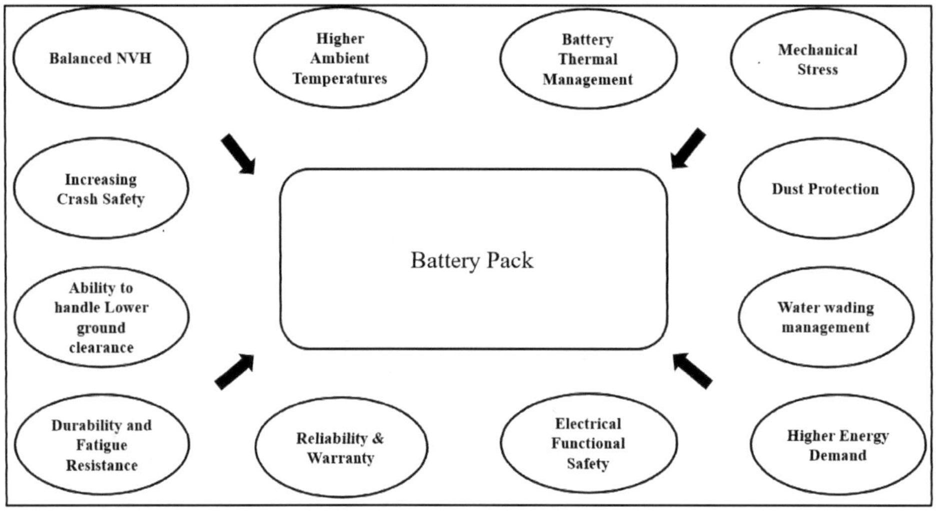

Figure 3. Key factors influencing the battery pack location & size

2.2. Battery Cell Technologies – Current & Future

Battery technology is growing more rapidly than expected due to tough competition and developments in raw materials. Table.4. Shows the currently available chemistries for anode and cathode and their respective strengths and weakness [10], and Figure.4. shows the short to medium-term research and emerging technologies.

Table.4. Current lithium-ion battery chemistries [10]

Cell Chemistry	Energy Density	Volumetric Density	Safety	Thermal Stability	Cycle Life	Cost
LCO	High	High	Medium	Low	Medium	Medium
LMO	High	Low	High	High	Low	Low
NCA	High	High	Low	Medium	Medium	High
NMC	High	High	Medium	Medium	High	High
LFP	Low	Low	High	Medium	High	Medium
Graphite/Carbon	High	Low	High	High	High	Medium
LTO	Low	High	High	Medium	High	Low
Si	High	High	High	High	High	Low

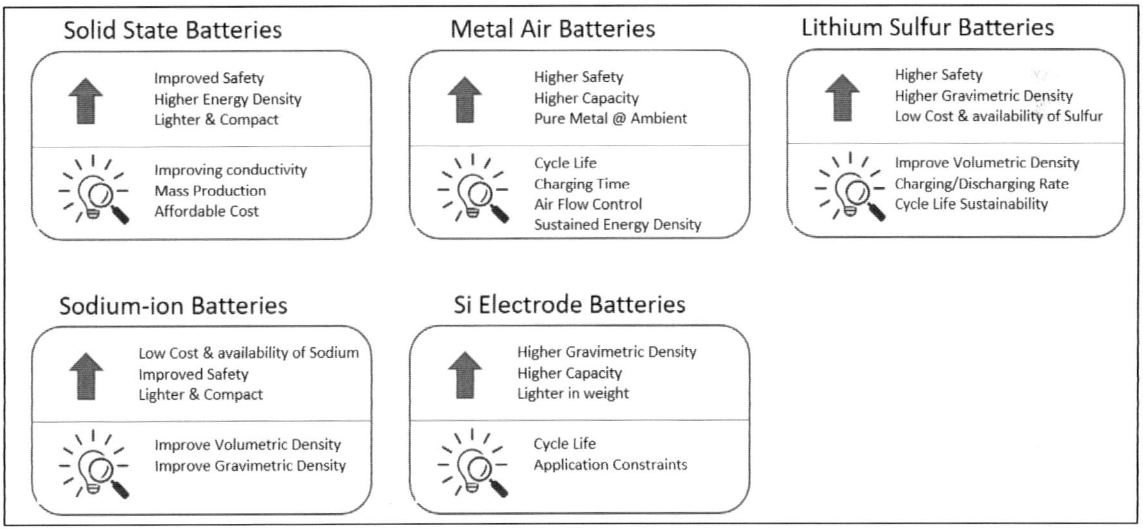

Figure.4. Promising battery chemistries: early-stage research [10]

Europe has invested heavily in research, manufacturing active materials, applications, and recycling [11]. China is leading EV battery pack manufacturing, followed by Europe. In terms of mobility space application, it is dominated by USA, UK, and China markets.

2.3. Testing & Validation of Battery

Post the battery integration into the vehicle platform; many tests are performed to qualify the battery pack and vehicle [12]. Some of the EV testing details are given below in Table.5. The Battery Pack Requirements & Integration are essential [13] in deciding the packaging location, space available, packaging efficiency, and the required Form factor. In addition, it must be integrated with the entire powertrain system and vehicle for seamless energy transfer. Therefore, it is the specific reason for choosing the correct form factor and potential standardization possible. Still, it is probably the most significant challenge in the development of battery packs [14].

IOP Conf. Series: Earth and Environmental Science 1042 (2022) 012012 doi:10.1088/1755-1315/1042/1/012012

Table.5. EV testing standards for the Indian market.

Category	Standard Used	Description
EV	AIS 038	Part I: Requirements of a vehicle with regard to its electrical safety
EV	AIS 039	Measurement of Electrical Energy Consumption
EV	AIS 040	Method of Measuring the Range
EV	AIS 041	Measurement of Net Power and the Max.30 Min.Power
EV	AIS 049	CMVR Type Approval for Electric Power Train Vehicles.
HEV	AIS 102 (I &II)	CMVR Type Approval for Hybrid Electric Vehicles
System	AIS 123 (I, II &III)	CMVR Type Approval of Hybrid Electric System Intended for Retro-fitment on Vehicles
Battery	AIS 048	Battery Operated Vehicles - Safety Requirements of Traction Batteries.
EVSE	AIS 138 (I & II)	Electric Vehicle Conductive AC /DC Charging System

3. R&D in Potential Battery Chemistries and Technologies

Many researchers are working on various innovations, materials, packaging, and cost reduction areas in EV batteries in today's research domain. It is a continuous process, as the technologies evolve every year, and many OEMs are gearing up for EV mobility. Any EV battery that proves outright from the material selection, concept proof of mass production adoption will nearly take about ten years [10], as given below using the research process gate as shown in Figure.5.

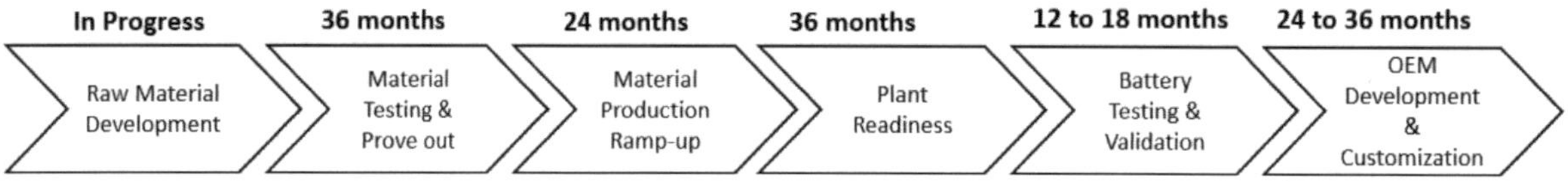

Figure 5. Research process gate for battery development

Table.6. shows the key areas where batteries should be in 20 years.

Table.6. Critical areas of EV battery development forecast [10]

Parameters	Now	2035
Cost	\$130/kWh (Cell)	\$50/kWh (Cell)
	\$280/kWh (Pack)	\$100/kWh (Pack)
Energy Density	700 Wh/l, 250 Wh/kg (Cell)	1400 Wh/l, 500 Wh/kg (Cell)
Power Density	3 kW/kg (Pack)	12 kW/kg (Pack)
Safety	Thermal Runaway	Elimination at Pack level
1st Life	8 Years (Pack)	15 Years (Pack)
Temperature	-20°C to +60°C (Cell)	-40°C to +80°C (Cell)
Predictability	Limited Performance Prediction Models	Full Predictive Models for Performance & Aging
Recyclability	10 to 15 % (Pack)	95 % (Pack)

4. Battery Supply-demand Vs. cost

To have a sustained and competitive electrification journey for the automakers, it is essential to have a robust and flexible battery supply chain. It starts with adequate access to battery raw materials like lithium, cobalt, manganese without any trade restrictions. It is also essential for automakers and suppliers to come forward and assemble/manufacture lithium-ion batteries at their factories, near electric vehicles OEM factories to avoid higher logistics and inventory costs. Due to the increased use of batteries in an EV and energy storage applications, the demand increases, and huge price reduction was noticed in the last five years. By 2024, an average battery pack will cost around \$94/kWh and \$62/kWh by 2030 as per Bloomberg NEF [15], as shown in Figure.6.

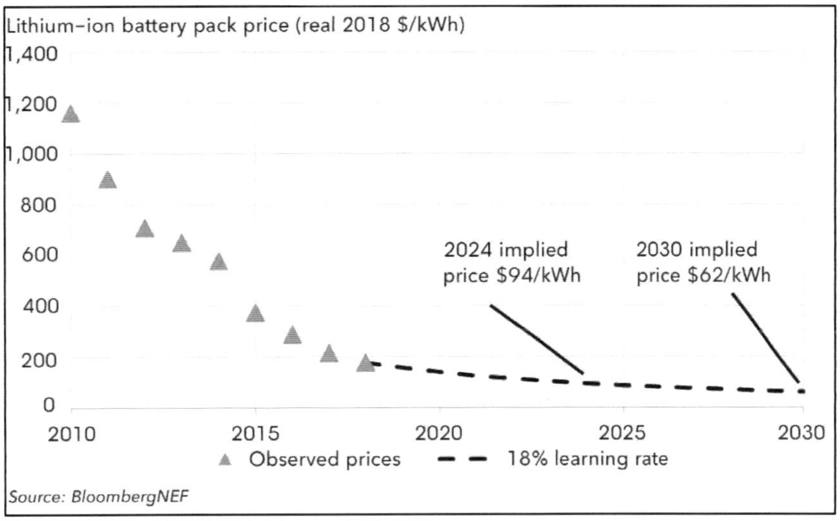

Figure 6. Lithium-ion battery price outlook [15]

The developments in the automobile industry to move towards green energy vehicles and in power generation industries markets to move away towards renewable energy sources are well underway [16]. However, lithium-ion batteries have a high dependency, advancing in lesser costs, energy density, and cycle life.

5. Stationary Applications

The Power demand and supply aspect is always very challenging, and a gap exists mainly during critical times. When we generate power from renewable energy sources, the generation pattern is unpredictable and highly time-dependent. This means we need a robust storage mechanism to store when energy is produced and release when there is a demand from energy consumers. This fluctuation in power generation and supply allows studying and researching robust storage technologies like battery storage systems[17]. When combined with battery-based energy storage, the net effect of variable renewable generation provides a steady-state generation that can be used in various applications. In addition, battery storage devices can compensate for the power fluctuation and ensure stability because of variable generation to sustain the grid stability. The Battery Storage Systems (BSS) were around 4 GW the year 2018 globally and it got doubled in the year 2020 as outlined in [18].

5.1. Battery Storage Systems & its functionality

Batteries store the electrical energy in the form of chemical energy. Lithium-ion based battery system is highly efficient in the order of 80% to 90%.

5.1.1. Lead Acid batteries. Conventional Lead-acid batteries have been widely accepted usage in the automotive industry for more than ten decades. It is primarily because of improved energy storage capacity developed over the years, and it has been used successfully in few grid applications and domestic power supply systems. The primary selection criteria are lower initial cost, higher depth of discharge, cycle life. These batteries are used in smaller PV installations. Careful disposal and recycling of used batteries is an important design requirement and research is in progress to recycle up to 100%. However, many challenges are involved due to poor infrastructure/industries' readiness for recycling and need a lot of commitment from key stakeholders. The critical element in the battery is lead, which needs to be extracted and used for battery manufacturing, leaving no hazard to the environment.

5.1.2. Nickel-based batteries. Ni-based batteries are used primarily when severe climate conditions like cold and require longer cycle time and faster-charging requirements. The usage of Ni batteries in energy storage applications is limited across the world. The advantage is its ability to work at a significantly lower temperature (up to -40 °C) due to its stable electrochemical properties. Due to its increased cycle life capacity quickly finds its usage in areas requiring a longer cycle at lower temperature conditions. Ni-Cd batteries are recyclable friendly, and with good infrastructure, nickel can be recovered. It is used for battery or allied industries, and cadmium can be used in the battery industry to avoid the need for new raw materials. We expect a shortage of nickel and cadmium in the following decades.

5.1.3. Sodium-based batteries. Sodium-Sulfur batteries are used in Stationary energy storage due to their efficiency, increased cycle life, higher energy density. These batteries require better temperature and electric protection, which demands better insulation cables to avoid overheating; hence, they are costlier than the other competing battery technologies. Sodium batteries in the early stages of commercialization of grid applications. R&D is still happening on the improved cycle life, specific energy density. When adapted to Industrial automation on a larger scale, the economy can be achieved. These batteries will find their usage due to higher operating voltage, performance on a good temperature range, better corrosion protection, and higher recycling ability [19].

Out of all the chemicals used for battery cell formulation, lithium is the preferred candidate due to its fabulous characteristics in storage properties, as below

- Safety
- Lightweight
- Longest Life
- Improved usage Capacity
- Constant Power
- Temperature Tolerant
- Supports Fast charging
- Improved calendar life
- Lower Maintenance
- Non-Hazardous

Lithium-ion batteries have grown in popularity because of their high energy and power densities. As the cost of the battery pack has fallen dramatically in the past decade, it finds its usage in both mobility and stationary energy storage. The point to be noted is the specification requirements for Li-ion batteries for these two applications vary due to the specific application & energy demand as shown in Table.7.

Table.7. Technical requirements of batteries for transportation and energy - storage applications [20]

Application	Energy & Power Density Requirements	Cycle Life	Required Lifetime	Safety	Efficiency	Discharge Rate
Transportation	High	500 (acceptable) 1000 (desired)	3 years	Some risk is acceptable	>85%	Medium
Grid-level Storage	Low to Variable	> 1000	10 to 20 years	Low	65 to 85%	Slow to Medium

Battery packs for mobility applications require compact size and require high volumetric energy and power densities. When Li-ion batteries are used for grid storage purposes [21], the form factor (both size and shape, weight) is not a major limiting factor as space availability is more. Whenever, the power generation & storage require an expansion, one more battery pack in the form of a container can be added to the existing packs. In the automotive application, the life of the Li-ion batteries varies from 6 to 8 years before its warranty and requires replacement [22]. In grid storage facilities, lithium battery packs can be used even up to 10 years.

Research and development in grid storage have developed many battery chemistries like lead-acid, nickel-cadmium, nickel-metal hydride (NiMH), and sodium-ion batteries [20]. But these chemistries failed to succeed in the market due to low take-rate and more in theoretical approach. For example, a 150 MW wind farm, in 2012 used the lead-acid installation for frequency regulation, but within five years, it was transitioned to lithium-ion batteries due to the superior performance of lithium-ion technology offers a longer life at a lower cost [23].

5.2. Importance of Battery Storage System for future (BSS)

According to the IEA 2019 reports [24], Wind energy and solar energy will contribute more than 50% of the total power generation by 2040. An interesting observation is power generation from coal (thermal power plants) is less than 3%. This indicates that by 2025, renewable sources will contribute to a more significant share of power generation than fossil fuels.

Figure.7. shows some exciting facts in a long-term scenario. The average annual energy investment will shift from conventional fossil fuels-based power generation to renewable energy sources from 2019 to 2050 [25]. This capital reallocation shows the increasing demand side and how investment can support power generation. The allocation of investment in renewable sources and battery storage is increasing from $610 billion to $1220 billion annually. The rising energy demand and concern on climate & greener production of power help in the rapid adoption of alternative or renewable energies. The use of batteries in stationary power storage will likely increase due to the gap between supply and demand requirements and the many advantages of batteries in electricity storage. The recent acceptance of renewable source-based power generation, improvement in the grid technology and minimizing the carbon footprint from conventional power plants, reduction in battery price is moving in the right direction to adopt the latest technologies in energy storage. This helps in efficient storage, distribution, minimal losses, and cleaner & greener means of energy storage.

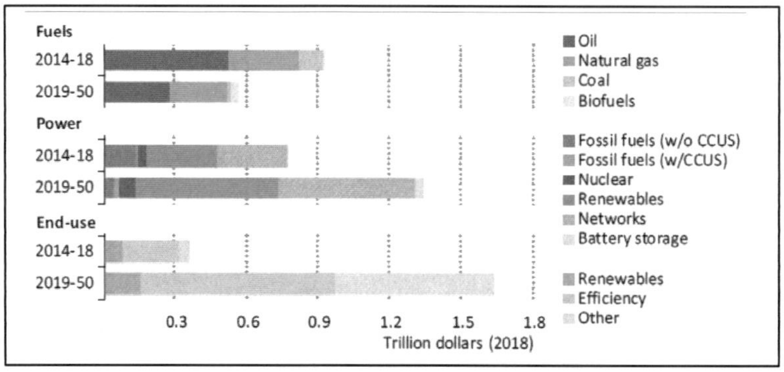

Figure 7. Average annual energy investment in sustainable scenario [24]

6. Discussion

This paper summarizes the two major application streams of Li-ion batteries, namely mobility and stationary applications. The design selection mainly in cell chemistry [26] and validation is a vital step primarily for mobility applications. The R&D in lithium ion is a key area and time consuming and many manufacturers & OEMs are investing huge cost & time. In BSS applications, the lithium ion is still a suitable candidate due to its increased cycle life and flexible charging/discharging capabilities. In India, the future energy generated from renewable energy sources is growing and naturally BSS will eventually a right choice for energy storage. The decreasing cost per kWh is a key indicator for the selection of battery technology for mobility and stationary applications.

7. Conclusion

The choice of Battery Storage system is considered primarily from the economic aspects like cost of storage, cost of operation, cost of maintenance. The recent technological growth and declining $ per kWh price of Li-ion batteries paved the way for increased use in mobility and storage applications. The OEMs and battery manufacturers are constantly working towards better cell chemistry, high energy density and make it more compact in size for improved integration into various applications. On-going research and development in areas like Solid state, sodium ion cells help in low-cost energy storage PowerGrid applications which is growing at a rapid rate. The challenges are robust battery thermal management systems, safety aspects, stringent testing protocols and certification standards to improve wider acceptance of battery storage for more application areas including air traffic, shipping, space travel.

References

[1] A.Yu, M.Sumangil, 2021 *Top* electric vehicle markets dominate lithium-ion battery capacity growth |*S&P Global Market Intelligence* (n.d.). Retrieved Januray 12, 2022, from https://www.spglobal.com/marketintelligence
[2] Y. Miao, P. Hynan, A.V. Jouanne, and A. Yokochi, 2019 Current Li-Ion Battery Technologies in Electric Vehicles and Opportunities for Advancements *Energies* 12, 1074, pp. 1-20
[3] R.Lindsay 2018 Why Lithium-Ion Batteries are the Future of UPS Energy Storage for Data Centers and Colocation Facilities, | *Data Center Frontier Special Report Series* (n.d.). Retrieved November 1, 2021, from https://datacenterfrontier.com/white-paper
[4] Dos Reis, G., Strange, C., Yadav, M., & Li, S. (2021). Lithium-ion battery data and where to find it. *Energy and AI*, *5*, 100081. https://doi.org/10.1016/J.EGYAI.2021.100081
[5] P.Tichelen ,,G.Mulder,A.Durand 2019 Preparatory Study on Eco-design and Energy Labelling of rechargeable electrochemical batteries with internal storage | *European Commission (n.d.).* Retrieved May 14, 2021, from https://ecodesignbatteries.eu/

[6] N.Niese, C.Pieper, A.Arora, and A.Xie 2020 The Case for a Circular Economy in Electric Vehicle Batteries | *BCG*. (n.d.). Retrieved July 6, 2021, from https://www.bcg.com/en-in/publications/2020/case-for-circular-economy-in-electric-vehicle-batteries
[7] Kurt Baes 2018 Future of batteries | *Arthur D. Little*. (n.d.). Retrieved August 18, 2021, from https://www.adlittle.com/en/insights/viewpoints/future-batteries
[8] D.Merriman 2018 The EV Revolution: Impacts on critical raw material supply chains| *Roskill Information Service (n.d.)*. Retrieved February 10, 2022, from https://www.minersoc.org/wp-content/uploads/2019/05/3ICM-Merriman.pdf
[9] Xu C, Dai Q, Gaines L, Hu M, Tukker A and Steubing B 2020 Future material demand for automotive lithium-based batteries *Commun. Mater.* **1**
[10] D.Greenwood 2018 AUTOMOTIVE BATTERIES 101| *Advanced Propulsion Systems,(n.d.)* Retrieved April 13, 2021, from https://warwick.ac.uk/fac/sci/wmg/
[11] S.Bhattacharjya, N.Arora, S. Bakshi, G.Jasuja 2018 Towards Resource Efficient Electric Vehicle Sector in India| *The Energy and Resource Institute (TERI)*, Retrieved September 5, 2021, from https://www.teriin.org/
[12] Arora S, Shen W and Kapoor A 2016 Review of mechanical design and strategic placement technique of a robust battery pack for electric vehicles *Renew. Sustain. Energy Rev.* **60** 1319–31
[13] Arora S, Kapoor A and Shen W 2018 Application of Robust Design Methodology to Battery Packs for Electric Vehicles: Identification of Critical Technical Requirements for Modular Architecture *Batter. 2018, Vol. 4, Page 30* **4** 30
[14] Löbberding H, Wessel S, Offermanns C, Kehrer M, Rother J, Heimes H and Kampker A 2020 From cell to battery system in BEVs: Analysis of system packing efficiency and cell types *World Electr. Veh. J.* 11 1–15
[15] L.Scot 2019 Behind the Scenes Take on Lithium-ion Battery Prices| *BloombergNEF*. Retrieved March 31, 2022, from https://about.bnef.com
[16] Ding Y, Cano Z P, Yu A, Lu J and Chen Z 2019 Automotive Li-Ion Batteries: Current Status and Future Perspectives *Electrochem. Energy Rev. 2019 21* **2** 1–28
[17] Trahey L, Brushett F R, Balsara N P, Ceder G, Cheng L, Chiang Y M, Hahn N T, Ingram B J, Minteer S D, Moore J S, Mueller K T, Nazar L F, Persson K A, Siegel D J, Xu K, Zavadil K R, Srinivasan V and Crabtree G W 2020 Energy storage emerging: A perspective from the Joint Center for Energy Storage Research *Proc. Natl. Acad. Sci. U. S. A.* **117** 12550–7
[18] Battery energy storage in the EU 2020| *Eurobat* Retrieved November 22, 2021, from http://www.eurobat.org/
[19] K T Smith 2021 What is the Future of Batteries for Stationary Energy Storage?| *AZoCleantech* Retrieved January 30, 2022, from https://www.azocleantech.com/
[20] H. Cooper, S.Brauer, D A Kalowb 2020 Battery Energy Storage in Stationary Applications, *Special Edition- Battery energy Storag*, Retrieved March 31, 2022, from https://www.aiche.org/
[21] Chen T, Jin Y, Lv H, Yang A, Liu M, Chen B, Xie Y and Chen Q 2020 Applications of Lithium-Ion Batteries in Grid-Scale Energy Storage Systems *Trans. Tianjin Univ.* **26** 208–17
[22] Masias A, Marcicki J and Paxton W A 2021 Opportunities and Challenges of Lithium Ion Batteries in Automotive Applications *ACS Energy Lett.* **6** 621–30
[23] Da Silva Lima L, Quartier M, Buchmayr A, Sanjuan-Delmás D, Laget H, Corbisier D, Mertens J and Dewulf J 2021 Life cycle assessment of lithium-ion batteries and vanadium redox flow batteries-based renewable energy storage systems *Sustain. Energy Technol. Assessments* **46**
[24] World Energy Outlook 2019| *IEA*, Retrieved August 18, 2021, from www.iea.org/reports
[25] AYuksel 2020 What is Stationary Energy storage an how energy storage works| *Cummins* Retrieved December 5, 2021, from https://www.cummins.com/news
[26] Beard, K.W., 2019. Linden's handbook of batteries. McGraw-Hill Education.

IOP Conf. Series: Earth and Environmental Science 1042 (2022) 012013 doi:10.1088/1755-1315/1042/1/012013

Exploring the state of solar photovoltaic decentralization in Ghana: trends and success factors

David M Sackey[1, *], De-Graft Owusu-Manu [2], Amevi Acakpovi [3], Frank K Appiah [4], David J Edwards [5] and Reza M Hosseini [6,]

[1] Department of Electrical/Electronic Engineering, Ho Technical University, Ho, Ghana
[2] Department of Construction Technology and Management, Kwame Nkrumah University of Science and Technology, Kumasi, Ghana
[3] Department of Electrical/Electronic Engineering, Accra Technical University, Accra, Ghana
[4] Renewable Energy Division. Ghana Energy Commission, Accra, Ghana.
[5] School of Engineering and the Built Environment, Birmingham City University, Birmingham, UK
[6] School of Architecture and Built Environment, Deakin University, Geelong, Australia

*Email: davidsackey.co.uk@gmail.com

Abstract. This research explores the trends of photovoltaic (PV) decentralization in Ghana and provides an updated illuminating insight into the sector. A literature review was carried out to understand and map out the trends, risks, and success factors. Additionally, time-series analysis was used to explore the installation trends from 2013 to 2019. A one-sample t-test was used to test the significance of each of the underlying indicators. The study identifies 2015 and 2018 as the years with the highest PV installations. On regional distribution, the research identified Central (63.25%), Greater Accra (26.44%), and Upper East (3.99%) as areas with the highest cumulative rate of a PV installation, respectively. This research raises awareness on the emerging perspectives on PV investment risk decisions and success criteria in Ghana. Considerations for potential investment opportunities and the stage of PV installations are expounded. It brings to light the geographical spread of installation that will aid critical policy decisions. Findings would be invaluable in making profound changes in the state's policy formulation around PV. The study provides a sound basis for further research, to be directed towards complementing this work by focusing on the social impacts and further engagement of end-users with solar PV.

1. Introduction

There is an abundant of renewable energy (RE) in Ghana which include Hydro Solar PV, Biomass, and Wind [1]. The renewable energy master plan for the country has recently been revised. It has an objective to increase the portion of renewable from 42.5 MW in 2015 to 1363.63 MW by 2030 with grid connections amounting to 1094.63 MW within the same period. The plan intends to provide about 1000 off-grid communities with decentralized electricity[2].

According to [3][4], Ghana's power sector is faced with a myriad of challenges. Chief among all is the lack of proper investment resulting in dilapidated infrastructure, stemming from bad managerial decisions and political influence. As a result, the sector has not attracted adequate substantial capital investment for several decades. Electricity access has increased extensively in Ghana, given the cheap power price that is supplied from Akosombo Dam and the hydroelectric facilities (1.6 US Cents per kWh). This, in turn, resulted in low electricity tariff rates until the 1990s [5]. And demand outstripped supply, because of the country has move to a middle-income status as pointed out by [3], hence the introduction of thermal plants that run on fossil fuels.

Furthermore, additional energy will be crucial as the country attempts to end poverty; eradicate hunger; improve human health, and achieve the Sustainable Development Goals (SDGs). However, major challenges lie in finding sustainable ways to meet these energy demands with clean energy [6]. Amidst all these challenges and uncertainties, solar energy generation through PV appears to be the game-changer, with tariffs being more economical – compared to conventional generation methods [4][7]. Besides, PV has numerous benefits in terms of reducing carbon emissions [8].

According to the report by [9], solar PV technology has advanced rapidly globally in the last decade. The authors of paper [10] reported a global PV installed capacity of 94 Gigawatts (GW) in 2018. The paper [11] explored the need for changing the power matrix into a more solid, secure, proficient, and clean system. With these in mind, building integrated PV(BIV) are becoming ubiquitous[2], [7]. PV technology has become the preferred choice to provide cheaper electricity for isolated communities [12][13]. Within the off-grid communities, PV application is vital for health delivery, education, ICT, clean water etc [14]. This has created a stream of research allocated to the topic. Authors of paper [15] explored the possible economic viability of PV power in Ghana. [16] confirmed the findings of [15] that Governments' subsidies will boost investments in the industry with their study on historic trends to a sustainable economy. Similarly, authors of paper [17] investigated the state of reliability and degradation of almost two-decade old Polycrystalline modules in Ghana. In the works of [18], they fleshed out the latent factors that impede PV investment in Ghana and confirmed the findings by [19] and [16] that, subsidies and tax exemptions will boost PV investment in Ghana.

Despite the contribution of these research studies, none has explored the current PV installation trends, and therefore a current picture of the installation state of play remains elusive in Ghana. This research aims to complete the research gab. The intention is to critically explore the state of solar photovoltaic decentralization in Ghana. To this end, the following objectives are formulated. i)Analyze various PV system designs, ii) Ascertain the regional distribution of PV installations; iii) Examine the trend of solar PV installations from 2013 to 2019; iv) Explore the historical perspective of PV installations in Ghana.

This paragraph summarizes the structure of the this paper. A review of energy consumption through the lens of climate change is covered in the introduction. The historical perspective of PV dissemination in Ghana is outlined in section 2. Experimentation analysis are discussed in section 3. Section 4 examines results. Results discussions are done in section 5. In section 6, conclusions were done.

2. Materials and methods

2.1 Historical PV Perspectives in Ghana

According to [20] PV development has seen an exponential growth since its introduction from the 1950s in Ghana, and is still being pursued in the 1970s, after the oil crisis [21]. [22] confirm that religious missions have initiated many solar energy projects in Ghana since the 1980s. Initially, solar energy facilities were intended for elementary lighting in isolated villages, through several government initiatives with development partners. [21][23] confirm many of these installations are executed through donor support in a private-public partnership (PPP). They provided the learning curve as Ghanaians became innovative with solar systems [2].

In a research study by [24], the Government of Ghana adopted PV into the rural electrification plan, because it was almost impossible to provide electricity to island communities beyond the volta lake in the 1990s. However, the development of national policies on renewables can be traced back to 1983. At the end of 1991, an estimated total of 335 solar PV installations were done in Ghana amounting to about 160 kilowatts. [22] show an increasing trend in PV from 150 kW to 1000 kW from 1991 to 2003. In 2013, the largest solar PV power plant of 2 MW was commissioned in Ghana [19] By 2015, over 10 MWp PV installations can be counted in Ghana.

There have been several government PV projects. Notable among them include the Weichau project initiated in 1997 under a fee-for-service supported arrangement for the Isofoton and Respro projects. Some government initiatives like the National Electrification Scheme (NES), The Ministry of Mines and Energy led the off-gird electrification project that enable individuals and communities to finance PV [22]. Finally, the ARB Apex Bank Project, which involved the private sector in providing maintenance and system supply [2]. The Spanish/Ghana governments and UNDP/GEF RE service project in 1998 and 1999 respectively were some of the earliest major PV donor project. The sum of the budget for these two major projects accounted for around US$4.5 million[25].

Furthermore, significant progress has been made in diverse areas regarding solar PV applications, including the addition of radio and TV operations to basic lighting solar projects, led by the Dutch government in 2007. This was extended to vaccine refrigeration, PV water pumping for irrigation, street lighting etc.

According to [15], in their study of the potential and economic viability of PV power in Ghana, the government's intention to provide subsidies for the power sector will boost investment in the industry. [17] investigated the state of reliability and degradation of certain PV modules in Ghana to provide useful information to project developers and manufactures.[18] unraveled the latent factors impeding PV investment in Ghana and confirmed the findings by [19] and [16] that, subsidies and tax exemptions will boost PV investment in Ghana.

Together with the encouragement of off-grid schemes, numerous grid-connected schemes have also been developed. The 44 kW systems at Kwame Nkrumah University of Science and Technology (KNUST) are some of the leading projects in this regard [26]. Several public institutions and private industries have also adopted the technology and installed solar PV of different capacities, mainly to reduce their reliance on the national grid. Figure 1 shows a steady rise in this trend, as a result of the government policy, backed by a massive drive by Ghana's Energy Commission.

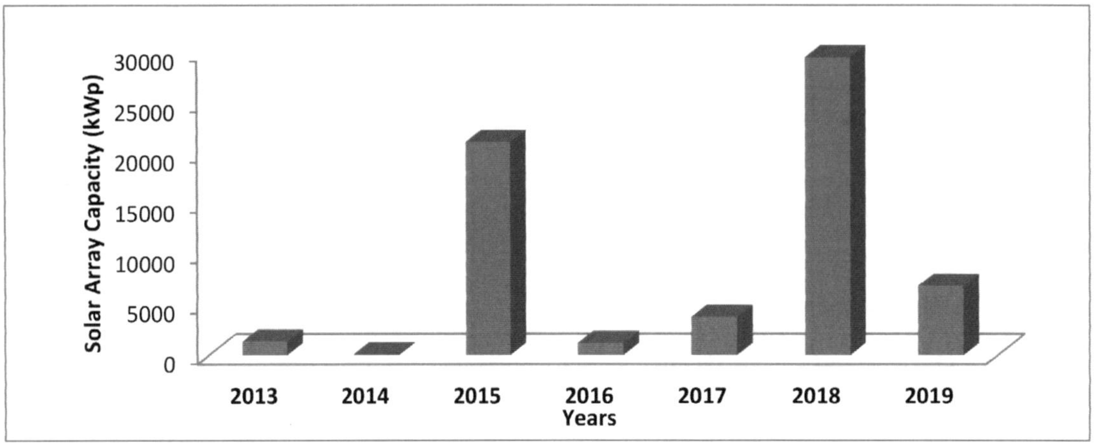

Figure 1. Yearly installation trend from 2013-2019-Capacity (kWp). **Source:** Author

2.2 Exploring the PV investment risk factors in Ghana

The PV infrastructure can be developed within 6-12 months, reasonably faster than alternative types of power plants that require over 4–5 years [27]. PVs nonetheless are becoming more attractive to utility providers, given that the costs associated with PV continue to fall [28]. Photovoltaic has revolved into the novel energy assets for the planet [11] and should be integrated into the grid to resolve the congestion on both distribution and transmission systems.

For any PV project to effectively take off, a preliminary conceptual outline ought to be created that assists to estimate capacity or megawatts (MW), desires, rough venture necessities, vitality yield, expected tax, and related income. Along these lines, business model evaluation can be made, including the degree of return on investment (ROI). An initial budget is normally created [29]

The dangers related to PV ventures are identified with the energy asset venture siting and allowing, solar-based innovation (generally new), potential debasement of PV modules. In Ghana, certain financing instruments such as venture capital, equity finance, debt financing, and crowd financing and many others are not fully developed [30][31]. In a research by [32], the weak Ghanaian cedi against many international trading currencies, predominantly the US Dollar (US$) and the rate of depreciation contributed to a rise in consumer price inflation [33]. Another risk metric that can be used to quantify the risks associated with an investment in solar PV plants is the Cash Flow at Risk (CFaR) [34].

In the works of [31], countless market-related risks such as immature supply chain, insignificant market size, and lack of fruitful PV project progress in Ghana. Market outlook and biases for example fossil fuel subsidies have rendered PV investment unappealing. Nonetheless, the development of renewable energy projects has seen significant improvement in recent times, possibly because of the presence of renewable energy legislation and policies in Ghana.

2.3 Experimentation Analysis

A deductive methodological approach was adopted for this paper. survey questionnaire was used to gather primary data. The survey help improves the consistency of interpretations and enhances replication. The research also depended on secondary data on PV installations in Ghana. The secondary source was from the Energy Commission of Ghana.

Undoubtedly, it may not be likely to find the precise population hence a theoretical sample was adopted. The sampling technique adopted in relation to its design, make inference is the purposive sampling. When a researcher wants to reach out to a targeted sample easily a purposive sampling is the most suitable[35]. Purposive sampling was adopted to reach the sample size of solar installation companies, government agencies, consultancy firms, and financial institution.

The two-page questionnaire was administered to eighty (80) respondents. Forty-seven (47) responses were received as in Table 1 below. The questionnaire was titled ‘Survey on exploring solar photovoltaic investment risk in Ghana.’ The questionnaire comprised two sections: A and B. Section A contains the demographic profile of the respondents that captured; the highest level of education, years of work experience with PV, and professional category. Section B comprised items to be rated on the level of their impact. The respondents were asked to rank PV investment risk based on a five-point Likert scale using the following Likert scale [1= Not a risk; 2= Somewhat of a risk; 3= Moderate risk; 4= Extreme risk; 5= Very Extreme risk] and tick (√) in the spaces provided.

The finalized set of questionnaires was distributed to respondents mainly from the public and private sectors who are experts with PV in Ghana. The questionnaire was distributed to respondents via Google form. The questionnaire was initially piloted with four prospective respondents before administering it. This was to minimize uncertainty, receive feedback, and finalize the questionnaire. [36], theorise the targeted respondent and, their ease of reading the survey instrument should be critically considered when drafting and organizing the survey instrument.

2.4 Respondents' profile

The respondents' highest level of education was found to be a Doctorate degree; having obtained a significant level of work experience mostly above 5 years. Their experience with solar photovoltaic will go a long way to enhance the quality of statistics to achieve the study objectives. The survey sourced information from technical groups (Utility Company e.g. VRA, ECG, Installation Company, Energy Commission). There were also investors (individual home users, financial institutions example banks, pension funds, insurance companies Ministry of Finance). Moreover, the last professional groups were policy institutions (NGOs, International Development Agencies like USAID, IFC Energy Commission). These respondents were chosen because [31] concluded they play diverse and consistent roles from policy design and implementation from local to national level which are crucial.

Table 1. Questionnaire response rate

Respondent job title	No.	%	Level of Education	No	%	Period of work (years)	No.	%
			Diploma	2	4.3	Below 5	2	4.3
			Master's degree		6.4	Below 5	2	4.3
Technical Experts	24	51.1		3		11 – 15	1	2.1
			Undergraduate Degree	19	40.4	Below 5	16	34.0
						6 – 10	3	6.4
			Master's degree	5	10.6	Below 5	4	8.5
						6 – 10	1	2.1
Individual Investors	17	36.2	Undergraduate Degree	9	19.1	Below 5	9	19.1
			Diploma	3	6.4	6 – 10	3	6.4
			Doctorate degree	1	2.1	6 – 10	1	2.1
			Master's degree	3	6.4	6 – 10	1	2.1
Policy Institutions (NGOs)	6	12.8				Below 5	2	4.3
			Undergraduate Degree	2	4.3	Below 5	2	4.3
Total	**47**	**100**		**47**	**100**		**47**	**100**

While the sample size is relatively small, it provides adequate coverage of technical experts based on the central limit theorem [37] and given that participants with extensive experience on the topic – 5 to 15 years – were involved. As argued by [38] and [31], assessing respondents' profile, particularly their extended professional practice are crucial features in establishing the authority of respondents in research studies.

Secondary data of time series on the solar array capacity (kWp) were used to analyze the installation trend from 2013-2019. These years mark the beginning of the renewable energy Act, licensing regime, and framework on government policies to utilize solar PV to provide electricity to off-grid communities and industries efforts to adopt the technology to meet their energy needs. The study incorporates trend analysis to investigate the situation of solar PV installation and distribution in Ghana. Descriptive statistics (mean score and standard deviation) were used and the results were graphically displayed using, frequency table, pie chart, bar chart, and line graph. Descriptive statistics enable the comprehension of massive amounts of data.

Given the increasing dispersed data on PV installation in Ghana, the analyses carefully focus on high-quality data provided by the Ghana Energy Commission (EC). The EC is the only institution in Ghana mandated by law to regulate, grant license, and supervise all providers of energy in the country. Data on 1920 installations across all the ten (10) regions in Ghana from 2013-2019 were analyzed. All retrieved data were pre-processed, and cleaned, to remove duplication and irrelevant information. The following fields were created in the data for the purposes of this study to aid the analysis: installation date, name of client/company, town, region, solar array capacity (kWp), system type, and type of mounting. The processed data were then aggregated into various categories and analyzed using spreadsheets and International Business Machines Statistical Package for Social Sciences version 26 (IBM SPSS). The SPSS tool was used to estimate the central tendency (sum) and descriptive analysis (a frequency with percent) and a sum of variables (a type of mount, capacity, and system type). Results including trends in frequency and percentage of each installation over time, the predominance of installation by region, and cumulative installed capacity (kWp) were graphically displayed using an excel spreadsheet. A similar analysis was performed according to the system type (grid-tied, grid-tied (non-net metered), grid-independent, home system, solar power street light, solar water pump system, standalone), and type of mount (carport, ground mount rooftop, etc).

3. Results and discussion

3.1 Reliability test

The research adopted several pointers to quantity the latent variables (constructs) – risk to solar PV investment factors. The reliability level of the pointers to the constructs was explored. Findings revealed that there is high internal consistency among measurement indicators since the Cronbach's Alpha value (0.861) was bigger than the cut-off point of 0.70. This indicates that the indicators measured their corresponding constructs, signifying a high level of reliability.

3.2 Connected installed PV according to different systems types.

The study analyzed the data based on the classification of installations in Ghana according to various categories of systems. From the data in Table 2, standalone systems have the largest number of installations amounting to 44834.57 kWp – this represents 75% of the total. Grid-tied (non-net metered) system is the second-largest, with an installed capacity of 11695.17 kWp, representing 13.9%.

Table 2. Connected PVs in Ghana by system type

System design	Frequency	Percent	Capacity
Grid Connected	128	6.7	5386.26
Grid tied (Non-Net Metered)	241	12.6	10129.91
Solar power street light	4	0.2	1364
Solar water pump system	14	0.7	47.30
Standalone	1450	75.6	2336.25
Utility-Scale	3	0.2	42500
Unknown system type	79	4.1	2088.91
Total	**1919**	**100.0**	**63852.63**

Regarding regional distribution, the cumulative installed capacity in Table 3 shows that the Central region has the highest share of installation with a total installed capacity of 40419.97 kWp, representing 63.25%. This is because the region hosts the biggest solar plant in currently the country with an installed capacity of 20kWp. Greater Accra region is the next region with a higher share of 16898.54 kWp – representing 26.44%. The next region with a high volume of installation is the Upper East region with a total installed capacity of 2553.38 kWp, which represents 3.99%. Volta region has the least installed capacity with a total installed capacity of 12.11 kWp representing 0.30%.

3.3 Classification of PV according to mounted design

Figure 2 presents various types of solar PV installations, segmented into ground-mounted, rooftop, carport, and some design categories. The study used descriptive statistics to present the total installations, followed by a percentage of installation and total installed capacity (kWp) for each category. As inferred from the pie chart in Figure 2, the ground mounting design has the largest share of PV installed, an installed capacity of 43176.62 kWp representing 68%. The rest are rooftop (20%), carport (11%), and unknown (1%).

Table 3. Combine regional PV installation by various system type

Region	System Type						Total install Capacity
	Grid Connected	**Grid tied (Non-Net Metered)**	**Utility-scale**	**Solar power street light**	**Solar water pump system**	**Standalone**	
Ashanti	15(16.5)	8(8.8)	-	-	-	68(74.7)	544.07
Brong Ahafo	-	-	-	-	1(11.1)	8(88.9)	48.2
Central	4(10.0)	14(35.0)	2(5.0)	1(2.5)	-	19(47.5)	40419.97
Eastern	1(1.8)	26(45.6)	-	0	-	30(52.6)	917
Greater Accra	106(6.8)	179(11.5)	-	1(0.10)	-	1266(81.6)	16885.8
Nationwide	-	-	-	2(100.0)	-	-	1320
Northern	1(12.5)	2(25.0)	-	-	1(12.5)	4(50.0)	631.7
Upper East	-	1(5.6)	1(5.6)	-	12(66.7)	4(22.2)	2553.38
Volta	-	8(21.6)	-	-	-	29(78.4)	12.11
Western	1(3.8)	3(11.5)	-	-	-	22(84.6)	154.94
Total	128(7.0)	241(13.1)	3(0.20)	4(0.2)	14(0.8)	1450(78.8)	365.45

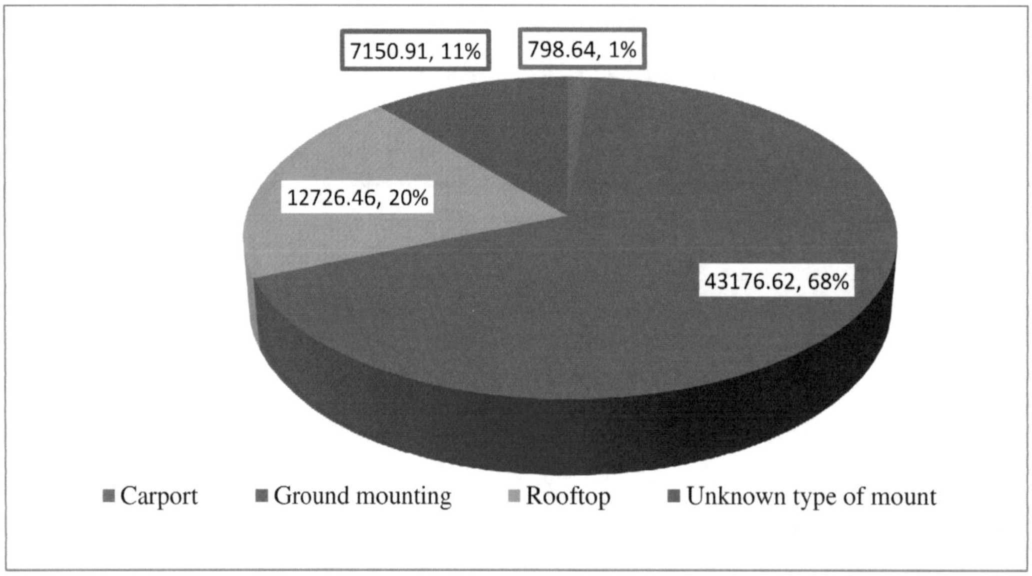

Figure 2. Connected PV in Ghana by type of mount

3.4 Regional installed PV distribution by capacity

In Figure 3, the regional distribution of PV installations is demonstrated, according to an installed capacity that exceeds 1000kWp from 2013 to 2019. This is important to differentiate between industrial and non-industrial users. Greater Accra, as the capital city, has a steady rise in installations from 2013 to 2019. The central region follows next. It has an intermittent installation that rose sharply in 2015 and slowed down in 2016. In 2017, installation began to rise gradually and peaked in 2018. The trend eventually deep in 2019. The upper east region is the third, having have its installation started between 2016 and 2017. The trend increased and picked in 2018 when it started to drop in 2019.

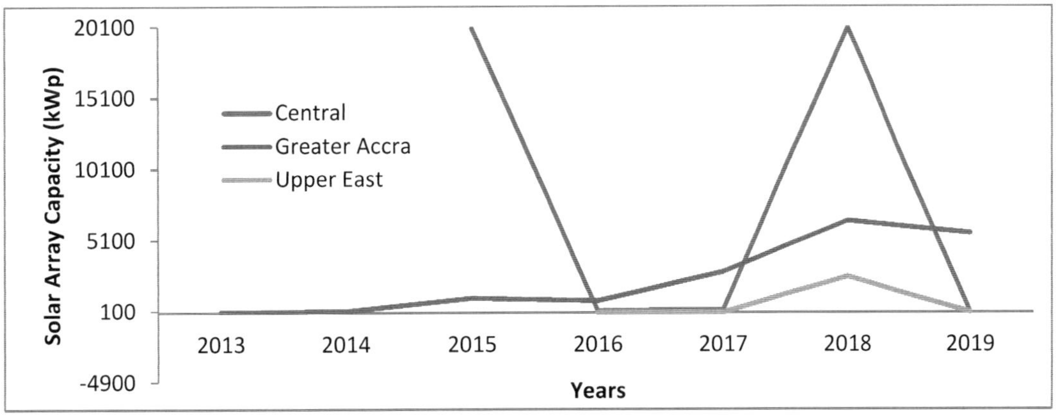

Figure 3. Regional PV distribution with capacity exceeding 1000kWp from 2013-2019

According to Figure 4, a regional distribution having installations less than 1000 kWp shows that there has been a gradual improvement of installation in eight (8) regions. The Ashanti region had a steady rise in installations from 2015 up until 2019. Followed by the Eastern region also has its fair share of the installations that started in 2015 and kept rising until 2019. The remaining regions had significant installed capacity, yet at intermittent levels.

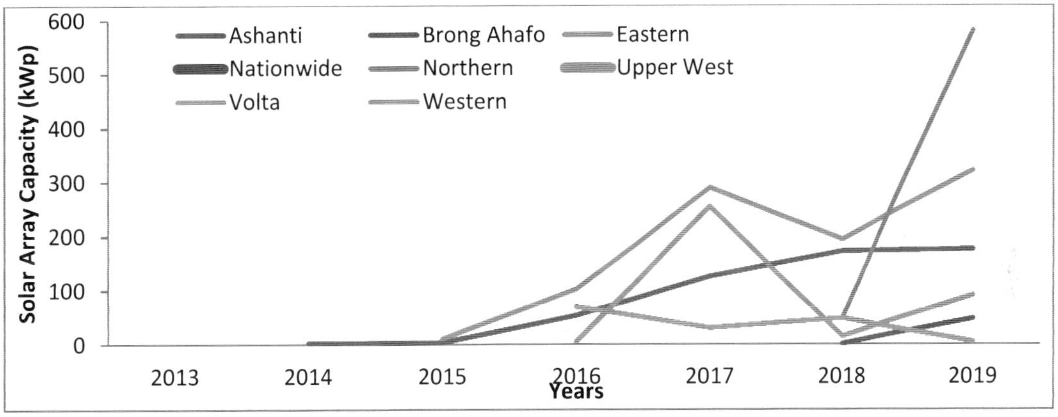

Figure 4. Regional PV distribution with a capacity below 1000 kWp from 2013-2019

4. Conclusion

This article explored the state of solar photovoltaic decentralization in Ghana from 2013 -2019. In doing so, the paper extended the work of [15] from 2013 to 2019 and shed light on human energy consumption through the lens of recent environmental changes on climate and public policy. Various historical perspectives of PV dissemination were also examined. Findings bring to light that in order to meet Ghana's industrialization objectives, reduce energy poverty and meet the SDGs [39], the exploitation of PV is vital, especially in the provision of electricity for deprived and island communities. This will help alleviate the perennial energy challenge in the country. Hitherto, a clear policy framework and effective implementation roadmap have remained missing in Ghana's RE vision [21]. That said, the Government of Ghana has recently shown clear policy commitment by the enactment of the country's renewable energy Act 835 backed by various interventions This includes the renewable energy technology transfer project, scraping of import duties on certain PV accessories,

creating a conducive environment for the technology transfer, implementation of capital subsidy for rooftop solar PV programme, and Sustainable Energy for All (SE4ALL) initiative just to name a few. This has resulted in a massive public and private investment within the PV industry in recent times.
PV installations and market penetration originate from conducive environment, for example the feed-in-tariffs (FiT), decent financing prospects, presence of skilled PV labour, and public acceptance of the technology[25]. Admittedly, this research identifies the major investment risk factors to be financing strategy, currency fluctuations, market obstructions, and regulatory framework and support mechanisms. This supports the argument by [40][31] [17][41]. The paper argued that addressing these factors would accelerate the decentralization of the technology and help to meet the governments RE policy agenda for 2030.

As other contributions of the study, six categories of PV installations designs, and four mounting types were identified. Findings also show that three regions have the highest number of installations. Policy analysis, business modeling, and geographically spread of the technology are the key implications of this research. The trend analysis of the installations shed light on some valuable understandings of the geographical spread of the PV technology at the regional levels. That is, the research identified the urban centers led by the Central region having the biggest installed capacity of PV installations. Energy poverty is predominant in rural villages and with this trend, it will be difficult to achieve universal energy access to meet the SDGs. The recommendations below are

- The government must encourage PV dissemination into deprived communities. This will enable the government to provide universal electricity access to its citizens.
- Governments support to financial institutions to enable them provide flexible long term financial backing to individuals and industries to be able to afford the technology. This will help meet our energy demands, prevent perennial energy crises, and cut down on the country's CO2 emission to improve the quality of life.

The study revealed a steady rise in both on-grid and off-grid installation year-on-year. On regional dissemination, the Central region has the largest share of PV installation followed by Greater Accra and Upper East respectively. With these findings, the research contributes to enhancing the awareness of global investors, entrepreneurs and financial institutions of PV project, conditions and investment that stimulate the growth and dissemination of PV in Ghana. From a practical perspective, this research provides a guide to private investors and the government to develop policies on stimulating RE investment.

This article has combined time-series data and literature to contribute to filling the theory gap on the state of solar PV dissemination, and its contribution to power generation in Ghana. In doing so, it has shed light on the financing obstacles in the sector. The findings demonstrate that there has been a massive increase in the rate of public and private PV investment in recent times compared to the early 2000s. Evidence from this paper has proved that the market is fully developed for commercial usage of solar PV. The paper proves a quantum leap of PV adoption in both individual homes, industries, and public institutions. It further suggests that, if this trend should continue at this rate, government will meet its RE target of 1094.63 MW by 2030.

This study is constrained to the PV industry in Ghana. Future research design to analyse the impact of PV adoption by private businesses, individuals, and on the national grid will significantly add to this knowledge gap.

References

[1] S. Gyamfi, M. Modjinou, and S. Djordjevic, "Improving electricity supply security in Ghana - The potential of renewable energy," *Renewable and Sustainable Energy Reviews*, vol. 43. 2015, doi: 10.1016/j.rser.2014.11.102.
[2] 2019 Energy Commission, "Ghana Rnewable Energy Master Plan," Accra, 2019. [Online]. Available: http://www.energycom.gov.gh/files/Renewable-Energy-Masterplan-February-2019.pdf.
[3] F. A. Diawuo, A. Pina, P. C. Baptista, and C. A. Silva, "Energy efficiency deployment: A

pathway to sustainable electrification in Ghana," *J. Clean. Prod.*, vol. 186, pp. 544–557, 2018, doi: 10.1016/j.jclepro.2018.03.088.
[4] A. Acakpovi, G. Botwe, and D. M. Sackey, "Impact of energy efficiency and conservation programs on the national grid in some selected households in Ghana," *Energy Effic.*, vol. 15, no. 1, pp. 1–20, 2021, doi: 10.1007/s12053-021-09998-1.
[5] K. A. Agyarko, R. Opoku, and R. Van Buskirk, "Removing barriers and promoting demand-side energy efficiency in households in Sub-Saharan Africa: A case study in Ghana," *Energy Policy*, vol. 137, p. 111149, 2020, doi: 10.1016/j.enpol.2019.111149.
[6] C. Pradhan, M. K. Senapati, N. K. Ntiakoh, and R. K. Calay, "Roach Infestation Optimization MPPT Algorithm for Solar Photovoltaic System," *Electronics*, vol. 11, no. 6, p. 927, 2022, doi: https://doi.org/10.3390/ electronics11060927.
[7] J. D. Watson, N. R. Watson, D. Santos-Martin, A. R. Wood, S. Lemon, and A. J. V. Miller, "Impact of solar photovoltaics on the low-voltage distribution network in New Zealand," *IET Gener. Transm. Distrib.*, vol. 10, no. 1, 2016, doi: 10.1049/iet-gtd.2014.1076.
[8] F. Murphy and K. McDonnell, "A feasibility assessment of photovoltaic power systems in Ireland; a case study for the Dublin region," *Sustain.*, vol. 9, no. 2, 2017, doi: 10.3390/su9020302.
[9] J. Sawin, "Renewable Energy Policy Network for the 21st Century Renewables 2017 Global Status Report," 2017. [Online]. Available: http://www.ren21.net/wp-content/uploads/2017/06/17-8399_GSR_2017_Full_Report_0621_Opt.pdf%0Ahttp://dx.doi.org/10.1016/j.rser.2016.10.049%0Ahttp://www.ren21.net/status-of-renewables/global-status-report/.
[10] IEA, International Renewable Energy Agency, United Nations Statistics Division, The World Bank, and World Health Organization, "The Energy Progress Report," 2019. [Online]. Available: https://trackingsdg7.esmap.org/.
[11] R. Bayindir, I. Colak, G. Fulli, and K. Demirtas, "Smart grid technologies and applications," *Renewable and Sustainable Energy Reviews*, vol. 66. 2016, doi: 10.1016/j.rser.2016.08.002.
[12] R. H. Almeida and M. C. Brito, "A review of technical options for solar charging stations in Asia and Africa," *AIMS Energy*, vol. 3, no. 3, pp. 428–449, 2015, doi: 10.3934/energy.2015.3.428.
[13] L. G. P. V. S. U. Grid-forming *et al.*, "Active Power Control to Mitigate Frequency Deviations in," *Energies*, vol. 15, no. 6, p. 2035, 2022.
[14] A. Dajuma *et al.*, "Sensitivity of Solar Photovoltaic Panel Efficiency to Weather and Dust over West Africa: Comparative Experimental Study between Niamey (Niger) and Abidjan," *Comput. Water, Energy, Environ. Eng.*, vol. 5, no. 4, pp. 123–147, 2016, doi: 10.4236/cweee.2016.54012.
[15] S. Asumadu-Sarkodie and P. A. Owusu, "Energy Sources, Part A: Recovery, Utilization, and Environmental Effects The potential and economic viability of solar photovoltaic power in Ghana," *Energy Sources*, vol. 38, no. 5, 2016.
[16] J. Amankwah-Amoah and D. Sarpong, "Historical pathways to a green economy: The evolution and scaling-up of solar PV in Ghana, 1980-2010," *Technol. Forecast. Soc. Change*, vol. 102, 2016, doi: 10.1016/j.techfore.2015.02.017.
[17] D. Quansah, M. Adaramola, G. Takyi, and I. Edwin, "Reliability and Degradation of Solar PV Modules—Case Study of 19-Year-Old Polycrystalline Modules in Ghana," *Technologies*, vol. 5, no. 2, 2017, doi: 10.3390/technologies5020022.
[18] D. M. Sackey, D.-G. Owusu-Manu, R. O. Asiedu, and A. B. Jehuri, "Analysis of latent impeding factors to solar photovoltaic investments in Ghana," *Int. J. Energy Sect. Manag.*, vol. 14, no. 4, pp. 699–628, 2020, doi: 10.1108/IJESM-08-2019-0005.
[19] M. S. Adaramola, M. Agelin-Chaab, and S. S. Paul, "Analysis of hybrid energy systems for application in southern Ghana," *Energy Convers. Manag.*, vol. 88, 2014, doi: 10.1016/j.enconman.2014.08.029.
[20] S. Bawakyillenuo, "Effective government policy and institutional frameworks: Pivots for the dissemination and sustainability of solar PV in rural areas in the developing world: The case of Ghana," in *ISES Solar World Congress 2007, ISES 2007*, 2007, vol. 4, doi: 10.1007/978-3-540-

75997-3_580.
[21] K. G. Adanu, "Promoting photovoltaic electricity usage in developing countries - Experience from Ghana," *Sol. Energy Mater. Sol. Cells*, vol. 34, no. 1–4, 1994, doi: 10.1016/0927-0248(94)90025-6.
[22] G. Y. Obeng and H. Evers, "Solar PV rural electrification and energy-poverty: A review and conceptual framework with reference to Ghana," *ZEF Working Paper Series*, vol. 36, 2009.
[23] G. Y. Obeng, F. O. Akuffo, I. Braimah, H. D. Evers, and E. Mensah, "Impact of solar photovoltaic lighting on indoor air smoke in off-grid rural Ghana," *Energy Sustain. Dev.*, vol. 12, no. 1, 2008, doi: 10.1016/S0973-0826(08)60419-6.
[24] S. Bawakyillenuo, "Policy and institutional failures: Photovoltaic solar household system (PV/SHS) dissemination in Ghana," *Energy Environ.*, vol. 20, no. 6, 2009, doi: 10.1260/095830509789625383.
[25] S. Bawakyillenuo, "Deconstructing the dichotomies of solar photovoltaic (PV) dissemination trajectories in Ghana, Kenya and Zimbabwe from the 1960s to 2007," *Energy Policy*, vol. 49, 2012, doi: 10.1016/j.enpol.2012.06.042.
[26] M. Hossain, O. Al Farooque, M. A. Momin, and O. Almotairy, "Women in the boardroom and their impact on climate change related disclosure," *Soc. Responsib. J.*, vol. 13, no. 4, pp. 828–855, 2017, doi: 10.1108/SRJ-11-2016-0208.
[27] W. Muneer, K. Bhattacharya, and C. A. Cañizares, "Large-scale solar PV investment models, tools, and analysis: The Ontario case," *IEEE Trans. Power Syst.*, vol. 26, no. 4, 2011, doi: 10.1109/TPWRS.2011.2146796.
[28] Energia - The International Network on Gender & Energy, *Renewables 2018 Global Status Report (REN21) Released.* 2018.
[29] H. P. Janicki and P. V. Wunnava, "Determinants of foreign direct investment: Empirical evidence from EU accession candidates," *Appl. Econ.*, vol. 36, no. 5, 2004, doi: 10.1080/0003684041000l682214.
[30] Y. Ding, W. Shen, G. Levitin, P. Wang, L. Goel, and Q. Wu, "Economical evaluation of large-scale photovoltaic systems using Universal Generating Function techniques," *J. Mod. Power Syst. Clean Energy*, vol. 1, no. 2, 2013, doi: 10.1007/s40565-013-0017-z.
[31] E. Bensah, F. Kemausor, E. Antwi, and J. Ahiekpor, "Identification of Barriers to Renewable Energy Technology Transfer to Ghana," *Energy Comm. Ghana UNDP*, 2014.
[32] P. Alagidede and M. Ibrahim, "On the Causes and Effects of Exchange Rate Volatility on Economic Growth: Evidence from Ghana," *J. African Bus.*, vol. 18, no. 2, 2017, doi: 10.1080/15228916.2017.1247330.
[33] A. P. C. Chan, A. Darko, A. O. Olanipekun, and E. E. Ameyaw, "Critical barriers to green building technologies adoption in developing countries: The case of Ghana," *J. Clean. Prod.*, vol. 172, 2018, doi: 10.1016/j.jclepro.2017.10.235.
[34] B. Lu and M. Davison, "Analyzing the impact of environmental variables on the repayment time for solar farms under feed-in tariff," *Environ. Syst. Res.*, vol. 2, no. 1, 2013, doi: 10.1186/2193-2697-2-5.
[35] M. D. C. Tongco, "Purposive sampling as a tool for informant selection," *Ethnobot. Res. Appl.*, vol. 5, 2007, doi: 10.17348/era.5.0.147-158.
[36] R. O. Asiedu and H. W. Alfen, "ASSESSING THE FREQUENCY OF FACTORS ENGENDERING COST OVERRUNS USING FREQUENCY INDICES AND FACTOR ANALYSIS FOR PUBLIC BUILDING PROJECTS IN GHANA," *J. Int. Real Estate Constr. Stud.*, vol. 3, no. 2, 2013.
[37] B. G. Hwang, X. Zhao, Y. L. See, and Y. Zhong, "Addressing Risks in Green Retrofit Projects: The Case of Singapore," *Proj. Manag. J.*, vol. 46, no. 4, 2015, doi: 10.1002/pmj.21512.
[38] M. R. Hallowell and J. A. Gambatese, "Activity-Based Safety Risk Quantification for Concrete Formwork Construction," *J. Constr. Eng. Manag.*, vol. 135, no. 10, pp. 990–998, 2009, doi: 10.1061/(asce)co.1943-7862.0000071.

[39] D. Chirambo, "Towards the achievement of SDG 7 in sub-Saharan Africa: Creating synergies between Power Africa, Sustainable Energy for All and climate finance in-order to achieve universal energy access before 2030," *Renew. Sustain. Energy Rev.*, vol. 94, pp. 600–608, 2018, doi: 10.1016/j.rser.2018.06.025.
[40] J. Koinegg, T. Brudermann, A. Posch, and M. Mrotzek, "'It would be a shame if we did not take advantage of the spirit of the times ···': An analysis of prospects and barriers of Building integrated photovoltaics," *GAIA - Ecological Perspectives for Science and Society*, vol. 22, no. 1. 2013, doi: 10.14512/gaia.22.1.11.
[41] S. Ahmad, R. bin Mat Tahar, J. K. Cheng, and L. Yao, "Public acceptance of residential solar photovoltaic technology in Malaysia," *PSU Res. Rev.*, vol. 1, no. 3, 2017, doi: 10.1108/prr-11-2016-0009.

IOP Conf. Series: Earth and Environmental Science 1042 (2022) 012014 doi:10.1088/1755-1315/1042/1/012014

Optimal sizing and techno-economic analysis of a hybrid solar PV/wind/diesel generator system

Godfred Tay[1], Amevi Acakpovi[2], Patrick Adjei[1], George K. Aggrey[3], Robert Sowah[4], Daniel Kofi[1], Maxwell Afonope[1], Mustapha Sulley[1]

[1] Regional Maritime University, Department of Electrical and Electronic Engineering, Accra, Ghana
[2] Accra Technical University, Department of Electrical and Electronic Engineering, Accra, Ghana
[3] University of Cape-Coast, Department of Computer Science and Information Technology, Cape-Coast, Ghana
[4] University of Ghana, Department of Computer Engineering, Accra, Ghana

*Email: patrickadjei1@gmail.com

Abstract. Hybrid power systems that combine wind and solar PV technology have been widely employed for power generation, particularly for electrification in remote and islanding locations, because they are more cost-effective and reliable than traditional power systems. This article intends to develop an environmentally friendly and cost-effective hybrid power system for selected critical loads in the Avuto community of Ghana. Following the acquisition of site data, a hybrid solar PV, wind, diesel generator, and converter analysis was conducted using HOMER software to establish the appropriate sizing of system components based on technical and economic parameters such as load served, annual electricity production, net present cost (NPC), emission, Operating cost, Fuel consumption and energy cost (COE). Based on the optimization computational results, it can be stated that the combination of system components, including solar photovoltaic, wind turbine, and diesel generator, is a good fit for the application region and might be used for rural and island electrification in the future. The suggested energy system has an LCOE of 0.39 US$/kWh for the 1.08 US$/litre diesel fuel cost and a 3.33-year payback period, with 58.8 kW for PV, 7 units for 3 kW wind turbines, 10 kW for diesel generator, and 6.99 kW for the converter. In terms of emission reduction, the proposed case presented a 55% emission reduction from the base case scenario.

1. Introduction

In most developing countries such as Ghana, access to electricity for rural communities is challenging despite the pledge to ensure the United Nations sustainable development goal 7 (Access to reliable, clean and affordable energy to all). These rural communities are either isolated from the grid due to their geographical location or grid extension such areas are economically impossible. According to a study conducted by [1], the population of people without access to electricity are over one billion in the World. To address the rising need for electricity, some countries aim to extend the main grid to those places where power is scarce, while others rely on diesel generators to provide basic electrification when it is required. The former method, as previously stated, is difficult to implement because the grid extension to such areas is economically not feasible.

According to a World Bank study on rural electrification programs, each km of grid extension in challenging terrains like islands and forests might cost up to $22,000 [2]. On the other hand, several challenges remain for the later power supply solution, such as rising costs, fuel logistics and storage, and noise and carbon emissions in environmental pollution [3]. The cost of electricity generation might be more than three times the cost of the national electricity bill.

Biogas, Solar PV and Wind energy technologies have seen decades of researches leading to their technological development in recent times. For these reasons RETs can be applied with high potential and reasonable expenditure to develop this alternative energy system. However, there are still certain obstacles to overcome. Due to the unavailability of renewable energy resources, which are subject to weather conditions, it is difficult to adjust output power to satisfy load demand [3].

Electrification in some developing countries is based on diesel generator, which has some limitations, whilst using exclusively renewable sources for power generation is also challenged with several issues, including the intermittent nature renewable sources and their associated high cost of investment. The current trends for microgrids for isolated rural communities have emerged over the past few years to be an alternative and clean source of energy supply. As a result, hybrid renewable energy [4,5]. HRES plays an essential role in power supply due to its capability to improve reliability, cost efficiency, and environmental friendliness [6]. Renewable energy systems have numerous merits when hybridized to meet reliability of power supply throughout the day. When one renewable energy technology (RET) is unable to generate due some weather conditions, another RET can generate to supply the energy demand. For instance, when solar energy is not available during cloudy days, wind turbines or tidal energy can be used to supply the energy demand. The advantages of mixing different renewables for greater benefits are well explained by [7,8].

Similarly, for wind generation, the most important consideration is the location of the site, which receives a consistent amount of wind. This system is almost self-sufficient in non-conventional energy and decreases long-term energy prices, with a diesel generator hybridized with wind to be utilized as a backup in instances such as large loads or low renewable power availability [9,10].

A viable HRES study is developed in this study to find an optimal size of the Solar PV/Wind/Diesel generator system for improving the current electricity system on Avuto.

Small businesses linked to tourism services, such as souvenir shops, hotels and resorts and the fish market have recently had the potential to flourish as a result of the growth of tourists. Furthermore, the presence of tourists has drawn investment for the development of city infrastructure, buildings, hotels, and other amenities, thus raising the level of living in the local community. This development calls for effective and efficient telecommunication services to improve service delivery in this community. The objective of this study is to assess the techno-economic viability of introducing a hybrid renewable energy system to power a telecommunication base station and a tourism and community clinic in Avuto using HOMER software.

The absence of electricity supplied by engine-based diesel generators continues to plague rural areas where communications masts are built. The use of hybrid energy (PV/Wind/Generators) to power telecommunication towers can ensure energy continuity and ensure the telecommunication network in Avuto. Many researchers have conducted studies into the feasibility of supporting the energy demand of telecommunication base stations in remote areas in the world. This section presents studies that have been conducted in this special area.

In a comparative analysis study, [11] evaluated the economic and technical performance of standalone solar PV and diesel generator system to supply the power requirement of a telecommunication system in Malaysia. The authors used the HOMER software to design and compare the two systems. The technical and economic parameters used to evaluate these two systems were the life cycle cost, ability to meet the load requirement of the telecommunication system 24 hours each day. The results of the research highlighted the reduction in the net present cost of the proposed HRES to the conventional diesel generator power system from RM 610.639 to RM 421,244. Many similar studies, including [12–1]]relied on Homer to perform technical and economic analysis.

The authors in [18] conducted a study that assessed the financial and technical viability of different renewable energy technologies for telecommunications in India. Twenty-one locations in India were selected for this study. The study adopted LCOE for nine different hybrid renewable power configurations. The proposed configurations were technically evaluated using power supply availability for different hours (1hr, 2hrs, 4hrs, 8hrs, 12hr and 16hrs). The study revealed that among the nine configurations that were simulated, grid connected solar PV power system had the lowest LCOE for 1hr unavailability of power supply whiles another configuration with grid connected Wind, solar PV, Diesel generator and battery recorded low LCOE with 4hrs power unavailability.

Salisu et al. (2019) investigated a hybrid power system [19]. In this energy management system, different renewable energy systems were integrated to reduce the overall cost of hybrid power system. The An optimization method with multiple objective function aimed at minimizing cost, was adopted to select the best combination of riots' that will meet the load demand of the system. In addition, the study also reviewed other studies and their optimization models applied to similar RET configurations.

Again, [20] demonstrated with an FM transmitter station, the viability of a hybrid renewable energy system consisting of solar PVs and diesel generator system. The objective was to achieve a 100% reliability and reduction in cost of electricity. In Tunisia, [21] used a modelled a solar PV and wind hybrid system for Tunisia. The mathematical algorithm considered energy production, energy conversion efficiency, sustainability and resource maps of potential renewable sites. The algorithm used resources data to predict the hybrid renewable energy technology of the area.

Essalaimeh et al. (2013) used the weather resources of Jordan to evaluate the performance of a hybrid renewable energy system [22]. The experiment used variable load resistors to model different load conditions to measure their electrical energy productions. The HRES consisted of a solar PV and Wind technology. A simple payback period was estimated using mathematical modelling to assess the economic viability of the proposed hybrid system.

Kabalci et al. (2013) designed a renewable energy power plant using wind and solar PV system [23]. The authors adopted perturb and observed algorithm with a PI controller to control the proposed solar PV and wind energy system. Furthermore, in [24] a solar PV and wind energy design were applied to a mobile base station in a developing country to investigate the feasibility of the proposed standalone microgrid. The study was conducted in three different communities in DR Congo. HOMER computational software was used to design, simulate and optimize the solar PV and Wind energy system. Technical, economic and environmental impact assessment were conducted to rank the proposed system.

Using the HOMER computational software, Li et al. (2013) performed a technical and economic assessment of a standalone microgrid system using Solar PV, Wind and battery storage [25]. The proposed microgrid was designed and analysed using the results from the HOMER software. The simulation results of the study were 5-kW solar PV, 0ne 2.5kW wind turbine, 6.94 kWh unit batteries and a 5-kW converter for the configuration.

El Badawe et al. (2012) conducted a comparative analysis of a renewable energy system and a non-renewable energy system in Canada [26]. The authors adopted a telecommunication station as a case study energy consumer for the proposed comparative study. HOMER computational software was used to design, optimize and simulate the proposed configurations. The energy configuration that incorporated diesel generator was cost effective and reduced the running time of the diesel generator. This reduced the overall emission and fuel consumption of the base station under study.

The authors in [27] proposed an integrated hybrid RE power system that integrates an energy management system. The objective of the paper was to make the energy system efficient and again reduce the cost of energy. The system's effectiveness was tested using PSCAD platform.

Daud et al. (2012) designed and analysed a hybrid energy system to provide electricity for a small family in Palestine [28]. The RE resource availability and load requirement was used to model the viability of the proposed system. The authors developed a computer model to reduce the life-cycle cost and CO2 emissions of the proposed HRES for the family in Palestine. The authors discovered that employing hybrid systems to provide electricity to a residence in this community is effective. For the

IOP Conf. Series: Earth and Environmental Science 1042 (2022) 012014 doi:10.1088/1755-1315/1042/1/012014

isolated rural-urban community in Ghana, [29] used Homer software to compare wind/hydrogen fuel cell storage systems with wind/battery energy storage systems. The authors compared the NPC, LCOE, and autonomy days of the two storage systems.

A summary of literature suggests that hybrid renewable energy systems present a better option of providing electricity for remote and island communities at a lower cost and less environmental distraction as compared to grid extension or diesel generator alone. Unfortunately, few literatures have simulated and tested this hybrid renewable energy system to evaluate the potential advantages it presents in terms of cost of energy, ability to meet load demand, reduced investment cost and reliability for the selected location. The objective of this study is to conduct a technical and economic evaluation of a hybrid solar PV, Wind and diesel generator system for the selected community in Volta region of Ghana in terms of cost, availability and ability to meet the community's energy demand. This study has not been conducted before in the selected community and it presents an opportunity for stakeholders in the renewable energy sector and government to overcome the challenge of providing energy for remote and Island Communities at a lower cost but reliable.

2. Materials and Methods

This section seeks to explains to the methodology adopted to achieve the objectives of the study. This section presents the method for energy audit undertaken to determine the load profile of the study area. These sections explain the renewable energy resources, load profile, cost inputs, technical and economic parameters used to model and design the proposed system. In summary it describes how the HOMER computational tool was used to size and design the proposed hybrid system and the existing case. Figure 1 shows the methodology to achieve the objective of this study.

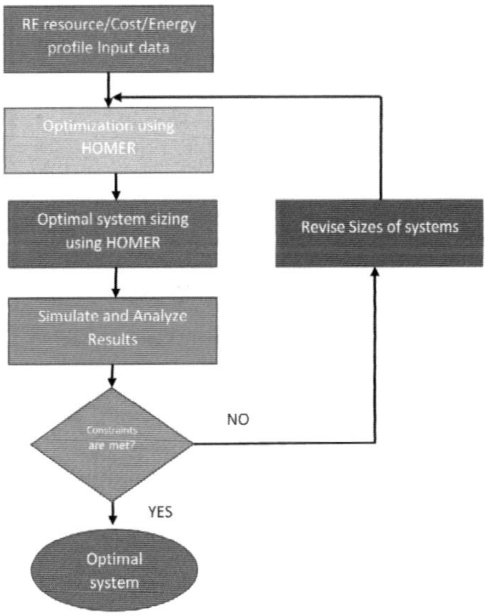

Figure 1. Methodology

2.1. *HOMER computational software*

HOMER computational software is adopted for the design, simulation, optimization and analysis of off-grid and grid-connected power system solutions for stand-alone, remote and distributed generations, both technically and financially. It allows the exploration of a wide range of technological solutions in order to account for energy resource availability and other factors. There are two optimization methods used by HOMER Pro. The grid search optimization technique replicates all of the Search Space's conceivable system configurations. The second and latest optimization method used by HOMER Pro is the propriety derivative-free method, which find the most cost-effective hybrid system whiles meeting the energy demand of the load. HOMER Pro will then show a list of configurations in order lowest net present cost (also known as life-cycle cost) for evaluation of the system design possibilities.

2.2. *Electric load*

The energy consumption profile of the study area is shown in Figures 2 (a) and (b). Figure 2 (a) represents the total primary load of the study which consist of the tourism office and the Community Hospital. Figure 2 (b) shows the telecommunication load on the 48V DC bus. The Telecommunication load remains constant because such services require a constant and uninterruptible power supply.

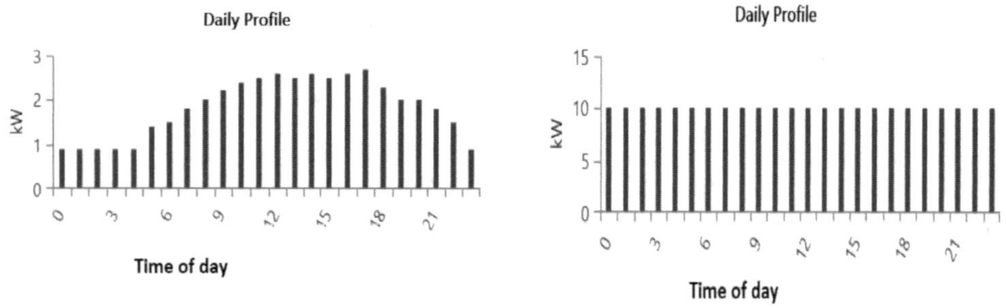

Figure 2. (a) load profile of Hospital and tourism centre (b) Telecommunication base station

2.3. *Geographical location of Avuto*

Avuto is a community in the Volta region of Ghana, the capital of Batanes, Avuto is about 69 mi or (110 km) North-East of Accra, the country's capital. Farming and fishing are the two main economic activities in this area. As indicated in Figure 3 (a) and (b), the community's location provides exceptional prospects for tourism.

(a) (b)

Figure 3. (a) location of Avuto (b) Location of Avuto on map

Table 1. Geographical location of the study area

Site	Longitude	Latitude	Altitude
Avuto	5.967543	0.711054	696m

2.4. Solar Resource of the study area

The solar radiation and clearness index of Avuto is shown in Figure 4. Avuto has average solar radiation of 5.52kWh/m^2/day with high radiations in February, March and January in order of highest radiation (6.09 kWh/m^2/day, 5.94kWh/m^2/day and 5.78kWh/m^2/day respectively).

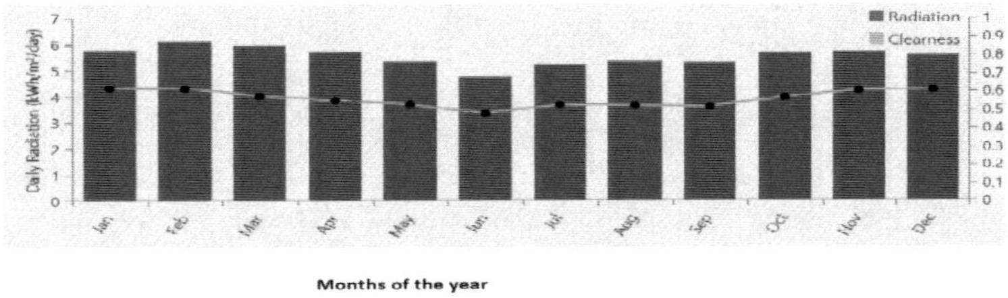

Figure 4. Solar radiation and Clearness index of Avuto

2.5. Wind resource of Avuto

The wind input data from the study area is presented in Figure 5. The wind resources were extracted from the National Aeronautics and Space Administration (NASA) weather satellite for Avuto. The average wind speed is 4.79 m/s with the highest winds recorded in July, August and September with a wind speed of 5.86 m/s, 5.93 m/s, and 5.46 m/s respectively. Other wind resource parameters considered for the analysis of the wind resources are presented in Table 2.

Table 2. Wind resource parameters

Parameter	Value
Anemometer Height (m)	30
Surface roughness	0.014
Weibull shape factor (k)	2

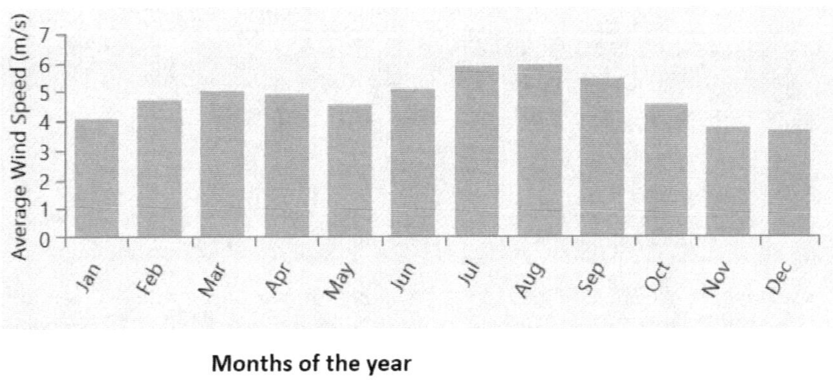

Figure 5. Wind speeds of Avuto

The proposed Solar PV/Wind/Diesel generator is designed in Homer software, simulated and optimized in Homer software. The design configuration is shown in figure 6. The proposed design incorporates two renewable energy technology into the existing diesel generator power system. The dispatch method for the proposed system is to have solar PV and wind serve the energy demand of the load and the diesel generator is used as a standby supply in cases where the former is not able to meet the load demand. The diesel generator provides power when the two renewable technology are not able to meet the load demand of the study area. This is to reduce diesel fuel usage and thereby reduce emissions.

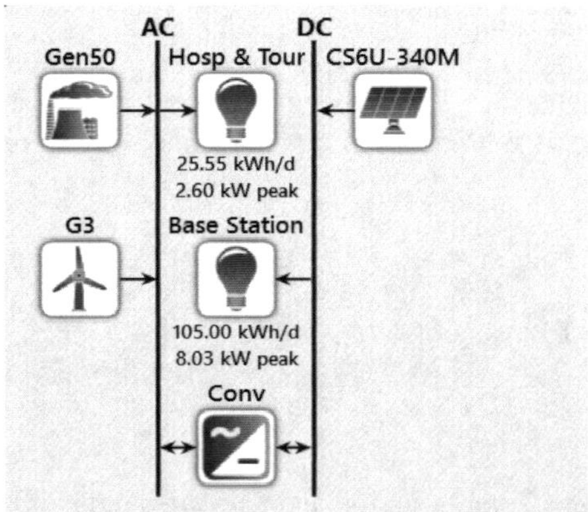

Figure 6. Homer design proposed system

2.6. *Solar PV Module*

The solar PV module with the technical and economic inputs used in this study are presented in Tables 3 and 4. The data used for the cost and technical specifications were adopted from the database of the HOMER software. These costs and technical input s from HOMER software reflect the average cost and characteristics of the solar PV used in this study.

Table 3. Cost input variables

Item	Cost
Initial cost ($/kW)	300.00
replacement cost ($/kW)	300.00
O & M cost ($/kW/year)	4.50

Table 4. Technical specifications of solar PV

Item	Specifications
Solar PV capacity (kW)	0.34
Operating temperature (^{0}C)	45
Efficiency (%)	17.49
Lifespan (Years)	25
Electrical Bus	DC

2.7. *Wind turbine*

The wind turbine with its associated cost and technical input variables are presented in Tables 5 and 6 below. Again, the data used for the selection of the wind turbine were collected from the HOMER software component library. These variables and their values are averagely similar to the market available wind turbines. The wind turbine used for the analysis is a low cost, a low-speed vertical axis wind turbine that is a commercially available off-the-shelve product.

Table 5. Cost input variables

Item	Cost
Initial cost ($/qty)	1885.42
Cost of replacement ($/qty)	1885.42
O & M cost ($/qty/year)	180.00

Table 6. Technical specifications of Wind turbine

Item	Specifications
Rated power (kW)	3
Cut-in-speed (m/s)	2
Rated speed (m/s)	15
Cut-out-speed (m/s)	20
Hub height (m)	30
Lifespan (Years)	20
Electrical Bus	AC

2.8. *Project economic input variables*

The economic input needed to analyze the proposed system economically are presented in Table 7. The economic input variable was adopted from the central bank of Ghana website. The project life of most renewable energy projects is 25 years [29–31]. Therefore a 25-year project lifespan was adopted to analyze the proposed system.

2.9. *Parameters for assessment*

The technical parameters adopted to assess this proposed system are Annual energy production (AEP), annual fuel consumption (AFC), total load served (LS) and Annual emission. These parameters were used because similar studies such as [29,32] just to mention a few used these parameters to technically assess a hybrid renewable energy system.
To calculate the annual energy production of the hybrid power system, equation (1) is used;

$$AEP = E_{PV} + E_{Wind} + E_{DG} \tag{1}$$

Where E_{PV} is annual electricity produced by the PV plant
E_{Wind} is the annual electricity produced by the wind farm
E_{DG} is the annual electricity produced by the Diesel generator

To estimate the total load served, the condition below must be fulfilled;
Annual electricity produced must greater or equal to the total load profile as shown in equation (2)

$$AEP \geq E_{Hosp\&Tour} + E_{Base} \tag{2}$$

$E_{Hosp\&Tour}$ is the annual energy consumed by the hospital and tourism centre
E_{Base} is the annual energy consumed by the base station.
The annual total emission is estimated using equation (3);

$$Em_{GHG} = Em_{CO_2} + Em_{H_2C} + Em_{PM} + Em_{SO_2} + Em_{N_2O} \tag{3}$$

Em_{GHG} is the GHG emissions
Em_{CO_2} is CO_2 emissions
Em_{H_2C} is H_2C emissions
Em_{PM} is Emission from particulate matter
Em_{SO_2} is SO_2 emissions
Em_{N_2O} is N_2O emissions

In addition, the Levelized Cost of Electricity (LCOE), Net Present Cost (NPC), and Operating Cost are used in this study to assess the proposed system's economic viability. Again, these parameters have been used by researchers to economically evaluate hybrid renewable energy systems [29,30,33,34]. The LCOE is estimated in this study using equation (4);

$$LCOE = \frac{TLCC}{TLC_{ey}} \tag{4}$$

$$TLCC = \sum_{t=1}^{n} \frac{I_t + O\&M_t}{(1+r)^t} \tag{5}$$

$$TLC_{ey} = \sum_{t=1}^{n} \frac{E_t}{(1+r)^t} \tag{6}$$

Where;
I_t is the investment cost in the time t;
$O\&M$ is the operations and maintenance cost in time t;
E_t is the energy generated in time t;
r is the discount rate
n is the Project life of the hybrid system.

The annual operating cost of the system is estimated using equation (7);

$$C_{oper,tot} = C_{ann,tot} - C_{ann,cap} \tag{7}$$

Where;
$C_{oper,tot}$ is annual Operating cost
$C_{ann,tot}$ is the annualized total cost ($/yr)
$C_{ann,cap}$ is the annual capital cost ($/yr)

2.10. Sensitivity analysis
In addition to the design and optimization of the proposed hybrid system, a sensitivity analysis was performed to assess the behaviour of the proposed system when the scaled annual average energy consumption per day is increased by 10% and 20%.

3. Results and Interpretations
This section presents the results of the simulation performed with the HOMER computational software. These results are presented in Tables, graphs and figures for interpretation. The section presents optimized results of proposed design, technical results, economic results, emission results and sensitivity results of the proposed design. The relevance of the results from the HOMER to the objective of the study are interpreted in this section.

3.1. Optimized results of the proposed design
The optimal sizing results of the proposed Solar PV/Wind/Diesel generator system is presented in Figure 7 from the Homer software. The optimal size of the proposed system consists of a 58.8kW of Solar PV, 7 wind turbines with each having a capacity of 3kW, a 6.99 kW converter and a 10-kW diesel generator.

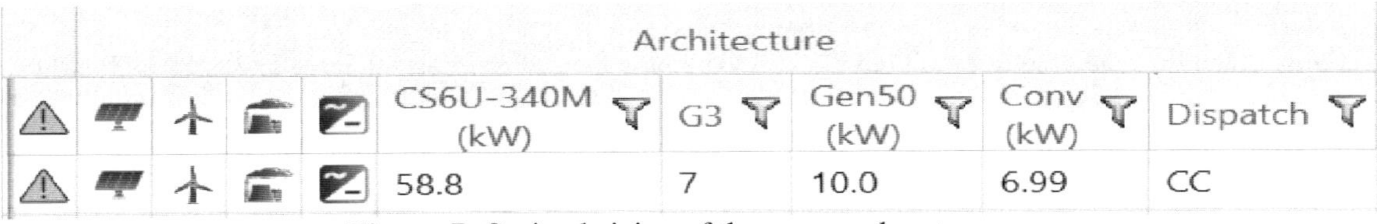

Architecture					
	CS6U-340M (kW)	G3	Gen50 (kW)	Conv (kW)	Dispatch
	58.8	7	10.0	6.99	CC

Figure 7. Optimal sizing of the proposed system.

3.2. *Technical results of the proposed design*

The parameters adopted to evaluate the technical viability of the proposed system with the results from the HOMER simulation of the proposed system using Homer software are presented in Table 7. The table shows results for AEP, AFC, LS and Emissions. In addition, the table compares the proposed case with the base case based on the technical parameters adopted in the study. The breakdown of electricity production by the solar PV/wind/Diesel generator system is presented in Figure 8. The figure shows the monthly electricity production by the proposed case.

Table 7. Technical results

Cases	Fuel Consumption (L)	AEP (kWh/yr)	Emission (Kg/yr)	Load served (kWh/yr)
Base Case	16,455	49,685	43,716	47,648
Proposed case	7,423	154,549	19,722	47,648

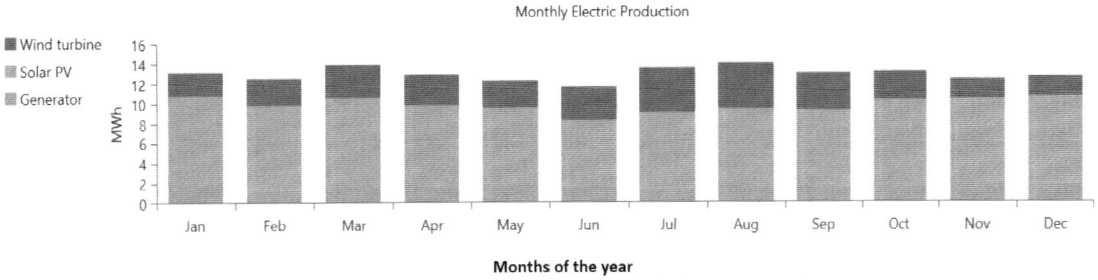

Figure 8. Detailed monthly production of the proposed system

3.3. *Economic results of the proposed system*

The results of the simulated proposed system are presented in Figure 9. The results show the costs obtained for the economic parameters adopted for the evaluation of the proposed system. The results also compare the proposed case with the base case concerning the economic parameters (NPC, LCOE and operating cost) used to evaluate them (See Table 8).

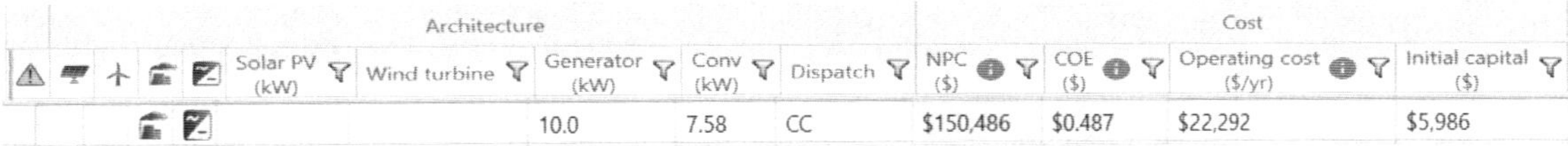

Architecture					Cost			
Solar PV (kW)	Wind turbine	Generator (kW)	Conv (kW)	Dispatch	NPC ($)	COE ($)	Operating cost ($/yr)	Initial capital ($)
		10.0	7.58	CC	$150,486	$0.487	$22,292	$5,986

Figure 9. Economic parameters

The table shows a lower LCOE for the proposed case as compared to the base case (Diesel generator alone). In general, the LCOE, NPC and Operating cost of the proposed case is lower than the base by 21%, 21% and 43% respectively.

Table 8. Economic comparison of cases

Cases	LCOE($/kWh)	NPC ($)	Operating Cost ($)
Base Case	0.49	150,486.00	22,292.00
Proposed case	0.39	118,788.00	12,658.00

3.4. *Emission results of base case and Proposed case*
Figure 10 shows the graph of total emissions of the base case and proposed case. The graphs also show the constituent of GHG emission of both base case and proposed case. These figures represent the emissions in kilogram per year.

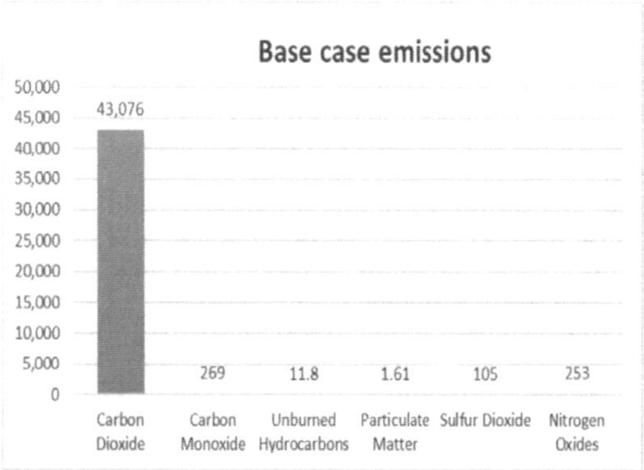

(a)

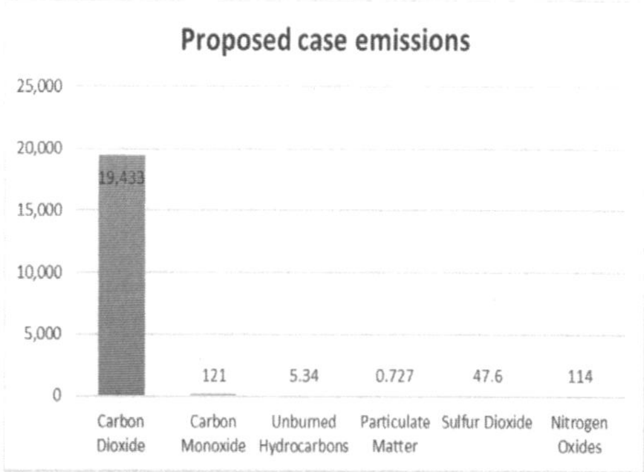

(b)

Figure10. (a) Proposed case emissions (b) Base case emissions

3.5. *Results for sensitivity analysis*
The results of sensitivity analysis performed on the proposed hybrid system using increasing scaled average energy demand are presented in Table 9 and 10. The sensitivity analysis evaluated the technical and economic impact of scaled annual daily energy demand.

Table 9. Technical results for sensitivity analysis

Scaled average load (kWh/day)	Fuel Consumption (L)	AEP (kWh/yr)	Emission (Kg/yr)	Load served (kWh/yr)
25.5 kWh/day	7,423	154,549	19,722	47,648
10%	7,580	152,761	19,844	48,557
20%	7,577	159,211	19,834	49,490
30%	7,635	163,186	19,988	163,186

Table 10. Economic results for sensitivity analysis

Scaled average load (kWh/day)	LCOE($/kWh)	NPC ($)	Operating Cost ($)
25.5 kWh/day	0.385	118,788.00	12,658.00
10%	0.381	119,807.00	12,872.00
20%	0.377	120,875.00	12,861.00
30%	0.373	122,040.00	12,938.00

4. Discussion
To meet the electricity demand of the load presented in this study whiles minimizing the LCOE, NPC and operating cost of the proposed system, an optimal size of solar PV, wind and converter was estimated Homer software. The optimal sizes estimated were 58.8 kW solar PV, 7 wind turbines (3kW each) and a 6.99 kW converter.

Furthermore, an AEP of 154,549kWh, which is 68% less than the base case (Diesel generator alone). The other technical parameters compared with the base case showed a reduced emission by 55%, and a reduction in fuel consumption by 55% as shown in Table 7. The proposed system served the electrical load 100% with annual excess electricity of about 68%. The annual unmet and capacity shortage for a system that has a renewable energy fraction of 58.7% were 0.0112% and 0.0914%. These figures compared with Single RE technology such as Solar PV and Wind alone are insignificant. This study is supported by previous studies such [35] which suggest that a hybrid system overcomes the limitation of a single energy source in terms of electricity production and reliability.

Moreover, the proposed case presented in this study reduced the annual fuel consumption by the diesel generator by 55% from the fuel consumption of the base case. As suggested by Wesly et al. (2020), electricity from diesel generators alone is expensive due to the high cost of fuel to remote and isolated areas. The authors suggested a hybrid system with a diesel generator in the mix to ensures reliability and reduces fuel consumption.

The proposed case recorded a total yearly greenhouse gas of 19,722 kg as compared to 43,716 kg by the base case. A significant annual emission reduction of 55%. A study conducted by [37] showed results that suggest that with a renewable fraction exceeding 72%, a hybrid system can reduce more than 2000kg of CO_2 emission annually. In this study, a renewable energy fraction of 58.7 % is reducing CO_2 emission of 23,643 kg from the base case. The results of this proposed system exceeded the projections of the study conducted by [37].

Again, the proposed case showed superior performance in terms of economic feasibility as compared to the base case. The LCOE of the proposed case was 0.39 $/kWh as compared to 0.49 $/kWh for the base case. The NPC and Operating costs of the proposed case were 21% and 43% lower than the base case economic performance. A study conducted by [38] concluded that a hybrid energy system is more reliable and cheaper than a single energy system.

The proposed hybrid solar PV, Wind and Diesel generator produced emission of 19,722 kg/yr of GHG whiles the base case produced 43716 kg/yr of GHG. From the HOMER simulation, it be deduced that the proposed case reduced the annual emission from the base case by 55%. Emission reduction is a major objective for the adoption of renewable energy. For this reason, a 55% reduction of emission from the proposed case is a significant achievement. This reduction presents the proposed case as environmentally feasible hybrid power system to serve the load requirement of the community. The sensitivity results show that an increase in the daily energy demand of the hospital and tourism centre increases the electric production of the proposed hybrid system. The increase also causes an increase in annual energy consumption and proportional increase in the annual emissions. The intermittent nature of renewable energy system requires a proportional increase in the energy production from diesel generator system. Again, an increase in daily energy consumption causes an increase in the net present cost and operating cost of the system whiles the levelized cost of energy slightly decreases.

5. Conclusion

This study was aimed at assessing the technical and economic feasibility of a solar PV/Wind/Diesel generator to serve a selected electrical load of Avuto community in the Volta region of Ghana. The electrical load considered in this study were a community clinic, a tourism centre and a telecommunication base station sited in Avuto.

A solar and wind resource of the study area NASA weather database. The average solar and wind resources with the highest and lowest values are presented. A proposed hybrid renewable energy system was designed using Homer software. The design was optimized to meet the electrical load whiles reducing, LCOE, NPC, Operating cost, fuel consumption and emissions. Again, the existing energy system which was termed in this study as a base case was also designed and simulated. The results of the optimized proposed case based on the economic parameters (LCOE, NPC and Operating cost) used were 0.39 $/kWh, $118,788.00 and $12,658.00 respectively. The base case on the other hand produced LCOE, NPC and operating costs of 0.49 $/kWh, $150,486.00 and $22,292.00 respectively. The technical parameters used for the comparison of the proposed case and base case were also presented in the study. The proposed case had AFC, AEP, Emission and LS to be 7,423L, 154,549 kWh/year, 19,722 kg/year and 47,648 kWh/year. For the same technical parameters, the base case had 16,455 L, 49,685 kWh/year, 43,716 kg/year and 47,648 kWh/year. The proposed case proved superior to the base case in both technical and economic analysis of the two cases. The study can confidently conclude that the proposed solar PV/Wind/Diesel generator is technically and economically feasible for implementation in the study area.

The study recommends the following for further investigation;

The proposed case in this current study had an excess electricity generation of 105,841kWh/year and a study to size a suitable energy storage system to store the excess electricity is critical.

A sensitivity analysis should be performed on the fuel price, average solar radiation, wind speeds and load consumption to test the robustness of the proposed case.

An economic assessment of the CO_2 emission reduction concerning carbon reduction credit to present its economic benefits.

References

[1] Tan Y, Meegahapola L and Muttaqi K M 2014 A review of technical challenges in planning and operation of remote area power supply systems *Renewable and Sustainable Energy Reviews* **38**

[2] Kazem H A, Albadi M H, Al-Waeli A H A, Al-Busaidi A H and Chaichan M T 2017 Techno-economic feasibility analysis of 1MW photovoltaic grid connected system in Oman *Case Studies in Thermal Engineering* **10** 131–41

[3] Ma T, Yang H and Lu L 2014 A feasibility study of a stand-alone hybrid solar-wind-battery system for a remote island *Applied Energy* **121**

[4] Ashourian M H, Cherati S M, Mohd Zin A A, Niknam N, Mokhtar A S and Anwari M 2013 Optimal green energy management for island resorts in Malaysia *Renewable Energy* **51**

[5] Haghighat Mamaghani A, Avella Escandon S A, Najafi B, Shirazi A and Rinaldi F 2016 Techno-economic feasibility of photovoltaic, wind, diesel and hybrid electrification systems for off-grid rural electrification in Colombia *Renewable Energy* **97**

[6] Muhammad Lawan S and Azlan Wan Zainal Abidin W 2020 A Review of Hybrid Renewable Energy Systems Based on Wind and Solar Energy: Modeling, Design and Optimization *Wind Solar Hybrid Renewable Energy System*

[7] Acakpovi A, Michael M B, Asabere N Y and Honvo J 2018 Modern electrical grid optimization with the integration of big data and artificial intelligence techniques *Advances in Energy Research* ed M J Acosta (New York, USA: Nova Science Publishers) pp 193–234

[8] Acakpovi A 2016 Original Framework for Optimizing Hybrid Energy Supply *Hindawi Publishing Corporation, Journal of Energy* **2016** 1–11

[9] Şen Z, Altunkaynak A and Erdik T 2012 Wind velocity vertical extrapolation by extended power law *Advances in Meteorology* **2012** 1–6

[10] Acakpovi A, Adjei P, Nwulu N and Asabere N Y 2020 Optimal Hybrid Renewable Energy System: A Comparative Study of Wind/Hydrogen/Fuel-Cell and Wind/Battery Storage ed P-F Pai *Journal of Electrical and Computer Engineering* **2020** 1–15

[11] Mohd Noor M S, Adzis Z, Arief Y Z and Muhamad N A 2016 Feasibility Analysis of Stand-Alone Renewable Energy Supply for Telecommunication Tower Using Homer *Applied Mechanics and Materials* **818**

[12] Acakpovi A, Michael M B and Majeed I B 2017 Grid connected hybrid solar and diesel generator set: A cost optimization with HOMER *Advances in Science, Technology and Engineering Systems* **2** 92–8

[13] Acakpovi A, ben Hagan E and Bennet Michael M 2015 Cost Benefit Analysis of Self-Optimized Hybrid Solar-Wind-Hydro Electrical Energy Supply as compared to HOMER Optimization *International Journal of Computer Applications* **114** 32–8

[14] Ajao K R, Oladosu O A and Popoola O T 2011 Using Homer Power Optimization Software For Cost Benefit Analysis Of Hybrid-Solar Power Generation Relative To Utility Cost In Nigeria *IJRRAS* **7**

[15] Acakpovi A, Hagan E ben and Michael M B 2015 Cost Benefit Analysis of Self-Optimized Hybrid Solar- Wind-Hydro Electrical Energy Supply as compared with HOMER Optimization *International Journal of Computer Applications* **114** 8887

[16] Lilienthal P and Gilman P 2004 *HOMER: The Micropower Optimization Model. National Renewable Energy Laboratory (NREL) Innovation for Our Energy Future (Fact Sheet)*

[17] Lal D K, Dash B B and Akella a. K 2011 Optimization of PV/Wind/Micro-Hydro/diesel hybrid power system in homer for the study area *International Journal on Electrical Engineering and Informatics* **3** 307–25

[18] Deevela N R, Singh B and Kandpal T C 2018 Load Profile of Telecom Towers and Potential Renewable Energy Power Supply Configurations *Proceedings of 2018 IEEE International Conference on Power Electronics, Drives and Energy Systems, PEDES 2018*

[19] Salisu S, Mustafa M W, Olatomiwa L and Mohammed O O 2019 Assessment of technical and economic feasibility for a hybrid PV-wind-diesel-battery energy system in a remote community of north central Nigeria *Alexandria Engineering Journal* **58** 1103–18

[20] Moghavvemi M, Ismail M S, Murali B, Yang S S, Attaran A and Moghavvemi S 2013 Development and optimization of a PV/diesel hybrid supply system for remote controlled commercial large scale FM transmitters *Energy Conversion and Management* **75**

[21] Colantoni A, Allegrini E, Boubaker K, Longo L, Di Giacinto S and Biondi P 2013 New insights for renewable energy hybrid photovoltaic/wind installations in Tunisia through a mathematical model *Energy Conversion and Management* **75**

[22] Essalaimeh S, Al-Salaymeh A and Abdullat Y 2013 Electrical production for domestic and industrial applications using hybrid PV-wind system *Energy Conversion and Management* **65**

[23] Kabalci E 2013 Design and analysis of a hybrid renewable energy plant with solar and wind power *Energy Conversion and Management* **72**

[24] Kusakana K and Vermaak H J 2013 Hybrid renewable power systems for mobile telephony base stations in developing countries *Renewable Energy* **51**

[25] Li C, Ge X, Zheng Y, Xu C, Ren Y, Song C and Yang C 2013 Techno-economic feasibility study of autonomous hybrid wind/PV/battery power system for a household in Urumqi, China *Energy* **55**

[26] El Badawe M, Iqbal T and Mann G K 2012 Optimization and a comparison between renewable and non-renewable energy systems for a telecommunication site *2012 25th IEEE Canadian Conference on Electrical and Computer Engineering: Vision for a Greener Future, CCECE 2012*

[27] Ghoddami H, Delghavi M B and Yazdani A 2012 An integrated wind-photovoltaic-battery system with reduced power-electronic interface and fast control for grid-tied and off-grid applications *Renewable Energy* **45**

[28] Daud A K and Ismail M S 2012 Design of isolated hybrid systems minimizing costs and pollutant emissions *Renewable Energy* **44**

[29] Acakpovi A, Adjei P, Nwulu N and Asabere N Y 2020 Optimal Hybrid Renewable Energy System: A Comparative Study of Wind/Hydrogen/Fuel-Cell and Wind/Battery Storage *Journal of Electrical and Computer Engineering*

[30] Acakpovi A, Adjei P, Asabere N Y, Sowah R A and Sackey D M 2021 Techno-Economic Evaluation of Hydrogen Fuel Cell Electricity Generation Based on Anloga (Ghana) Wind Regime *International Journal of Energy Optimization and Engineering*

[31] Xu X, Hu W, Cao D, Huang Q, Chen C and Chen Z 2020 Optimized sizing of a standalone PV-wind-hydropower station with pumped-storage installation hybrid energy system *Renewable Energy*

[32] Pal P and Mukherjee V 2021 Off-grid solar photovoltaic/hydrogen fuel cell system for renewable energy generation: An investigation based on techno-economic feasibility assessment for the application of end-user load demand in North-East India *Renewable and Sustainable Energy Reviews*

[33] Araoye T O, Ashigwuike E C, Egoigwe S V, Ilo F U, Adeyemi A C and Lawal R S 2021 Modeling, simulation, and optimization of biogas-diesel hybrid microgrid renewable energy system for electrification in rural area *IET Renewable Power Generation*
[34] Rezk H, Al-Dhaifallah M, Hassan Y B and Ziedan H A 2020 Optimization and Energy Management of Hybrid Photovoltaic-Diesel-Battery System to Pump and Desalinate Water at Isolated Regions *IEEE Access*
[35] Al-falahi M D A, Jayasinghe S D G and Enshaei H 2017 A review on recent size optimization methodologies for standalone solar and wind hybrid renewable energy system *Energy Conversion and Management* **143**
[36] Wesly J, Brasil A C P, Frate C A and Badibanga R K 2020 Techno-economic analysis of a PV-wind-battery for a remote community in Haiti *Case Studies in Chemical and Environmental Engineering* **2**
[37] Razmjoo A, Gakenia Kaigutha L, Vaziri Rad M A, Marzband M, Davarpanah A and Denai M 2021 A Technical analysis investigating energy sustainability utilizing reliable renewable energy sources to reduce CO2 emissions in a high potential area *Renewable Energy* **164**
[38] Abbassi A, Dami M A and Jemli M 2017 A statistical approach for hybrid energy storage system sizing based on capacity distributions in an autonomous PV/Wind power generation system *Renewable Energy* **103**

Numerical analysis using SolTrace software to determine the size and position of a central receiver for ganged heliostat system

Lalit Jadhav*, Prem Gadiya, Jwalandhar Girnar, Harshvardhan Bhusal, Prakash Gadhe

School of Mechanical Engineering, Dr. Vishwanath Karad, MIT-WPU, Pune, India

*Email: lalitjadhav28@gmail.com

Abstract. Concentrated solar power (CSP) technology has become a promising technique for harnessing the sun's energy. Ganged heliostat system, also called as central receiver system (CRS), a type of CSP system that uses a field of radially distributed heliostats (mirrors), which track the position of the sun in two axes and focus the sunlight on the centrally placed receiver. Heliostats are used to concentrate solar rays on the receiver, producing a huge amount of energy which can later get converted into electricity or heat. The receiver is kept at a sufficient height to absorb maximum reflected rays. In the present work, a numerical analysis method is used to optimize receiver design parameters such as receiver size, its position in horizontal and vertical planes, and to predict the heat flux intensity at the receiver surface. However, there is no direct method to decide these parameters and often results in a loss of time while optimizing them. In this research paper, a simple methodology has been proposed to design a central receiver for CRS consisting of ganged heliostat system using SolTrace 3.1.0 software. This method is effective in determining the size, the position of the receiver and heat flux distribution at the receiver surface in the initial phase of receiver system design.

1. Introduction

In the modern era, the field of renewable energy is expanding and flourishing everywhere. With the prospects of achieving a clean energy future, innovation in this field will help in improving efficacy, delivering products and reducing costs. Renewable energy, also called as clean energy, is derived from replenished natural resources. Renewable energy includes solar energy, wind energy, geothermal energy, biomass, and hydroelectric power [1]. For several years, solar energy has been harnessed for various purposes like growing crops, staying warm, and drying foods. It is a potent form of energy which can be used to heat, cool, and light homes and businesses. Today, we use the Sun's rays in different ways like power generation via solar cells, heating water, drying clothes, and so on. Some of the products that are based on the application of solar energy are solar collectors like heliostats, solar cells, solar cookers, solar water heaters, and solar furnaces [2]. Concentrating solar power systems (CSP) and the sun are used as heat sources in the new generation of power plants. Power tower systems, reflector dish systems, and linear concentrators are the three main types of concentrating solar power systems [3].

IOP Conf. Series: Earth and Environmental Science 1042 (2022) 012015 doi:10.1088/1755-1315/1042/1/012015

One type of concentrated solar power (CSP) technology is the central receiver system, also referred to as solar tower power. It is made up of a heliostat field (mirrors), a receiver in the centre, and a power block [4], [5]. This system employs a field of heliostats, each of which tracks the sun's position along two axes and directs the sun's rays to a receiver located at the top of the tower. The receiver is an integral part of the central receiver system, which has a high radiation absorption capacity and low thermal losses for all thermodynamic processes. It is situated on top of a tower at a point where reflected energy from heliostats can be intercepted most effectively [6]. It receives all reflected sunrays from the heliostat field and converts thermal energy for transmission to the heat transfer fluid (HTF). It is mainly classified as volumetric receivers, cavity receivers, and particle receivers [7], [8]. Volumetric receivers are generally made up of metal or porous wires. They used to produce a volumetric effect, which means that the irradiated side of the absorber is cooler than the medium leaving the absorber [9], [10]. The reflected radiation from the heliostats travels through an aperture into a box-like structure before colliding with the heat transfer surface in a cavity receiver [11], [12]. Particle receivers, the least researched type of solar receiver, use solid particles for absorbing focused direct solar irradiation, which serves as heat transfer fluid and storage medium [13].

Highly specialized resources are required to design and build efficient, cost-effective central receiver systems (CRS). A huge amount of capital is required to design, build and run these systems. The central receiver is an integral component of CRS and is used to evaluate its performance [14]. The receiver includes many complex parameters that need to be designed and optimized for better heat flux distribution. It is not practical to physically build and test the many possible preliminary prototypes as part of the receiver development path. Instead, computer codes or numerical analysis are used to model it and predict the performance. Design optimization can be accomplished much more efficiently this way, resulting in fewer prototypes and a shorter development path. Also, the flux distribution on the receiver surface highly depends on the receiver design. Therefore, numerical analysis is recommended to design the receiver parameters before its manufacturing. SolTrace is a Monte-Carlo ray-tracing software package, provided by National Renewable Energy Laboratory (NREL) [15]. It is specially designed to model concentrated solar power systems and new solar optical systems. A methodology is presented in this article to design the central receiver for ganged heliostat system using SolTrace 3.1.0. Different receiver parameters such as size, position in horizontal and vertical planes are defined, iterated and optimized to predict the heat flux intensity at the receiver surface.

Nomenclature

d	:	bottom diameter of inverted frustum, m
h	:	vertical height of inverted frustum, m
h_1	:	bottom diameter height from heliostat, m
h_2	:	upper diameter height from heliostat, m
s	:	slant height of inverted frustum, m
D	:	upper diameter of inverted frustum, m
DNI	:	direct normal irradiance, W/m^2
L	:	distance between target and heliostat (slant range), m
R_{1H}	:	horizontal component of ray1, m
R_{2H}	:	horizontal component of ray2, m
X	:	horizontal axis points towards west
Y	:	vertical axis points towards zenith
Z	:	horizontal axis points towards north
α	:	semi-cone angle, °
β	:	ray inclination angle, °
θ	:	taper angle of inverted frustum, °
σ	:	Standard deviation

Subscripts and Superscripts

a	:	aim point unit vector
s	:	sun unit vector
t	:	target unit vector
A	:	aim point position
H	:	heliostat position
T	:	target coordinates

2. Ganged Heliostat System

Ganged heliostat system is a small scaled version of a solar power tower, developed and patented by Patwardhan et al [16]. A schematic view of the generated CAD model of the ganged heliostat system with the cylindrical receiver is shown in figure 1. It includes a heliostat field and a central receiver mounted on a rotating platform. There are a total of sixteen heliostats (mirrors) of size 0.37 x 0.37 m each. These mirrors are evenly distributed over four rows or actuating members with a common positioning mechanism to concentrate sun rays on the receiver. Sun's position is tracked in two axes simultaneously. The rotating platform tracks the sun's azimuthal position and heliostats track the sun's altitude position (elevational tracking). A primary DC motor (0.25 hp, PMDC Geared) and sprocket and chain mechanism are used to rotate the rotating platform. Two small DC motors (50 kg/cm) coupled with encoders are used to move actuating members which in turn move heliostats. A 3D CAD model of the system is developed in SolidWorks 2016 software. Sunrays intersections (both incident and reflected rays) are also depicted in the figure to understand the working of the system. The receiver which is going to be modelled in the latter part of this research paper is designed for ganged heliostat system to check its performance. Various required parameters such as size and position of each heliostat with respect to the centre, etc are referred from data available in [16]. This referred data is further used for developing different SolTrace models.

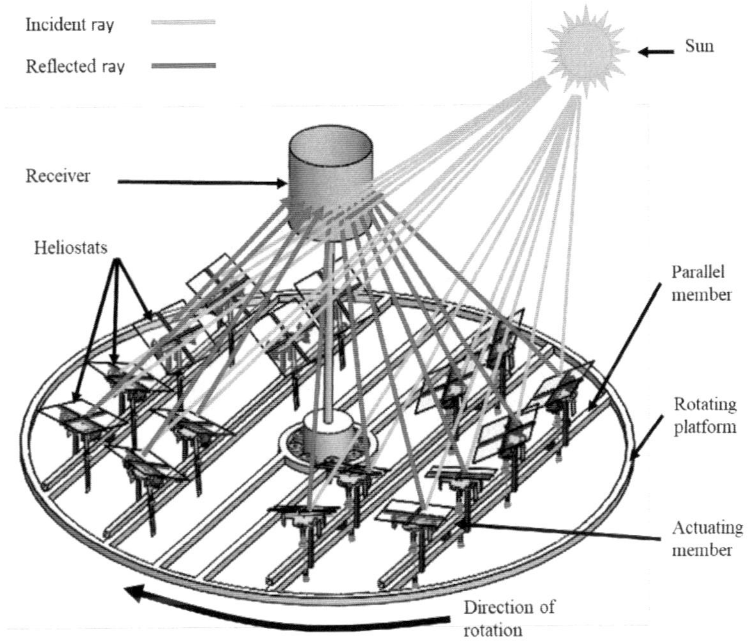

Figure 1. Isometric view of CAD model of ganged heliostat system with ray intersections

3. Methodology

In the present ganged heliostat system, sixteen heliostats track position of the sun and focus sunrays on the centrally placed receiver. Due to the circular arrangement of the heliostat field, the receiver receives sun-rays from all directions. To concentrate maximum reflected rays from the heliostats, central receivers are generally designed in a cylindrical shape and are mounted at a particular height from the heliostat field. Therefore, central receiver design includes the determination of receiver size and position from the ground. The same methodology has been followed in this paper. Heliostat data (aperture, nature of the surface, and position in the heliostat field) are taken from ganged heliostat system and modelled in SolTrace 3.1.0 software. Various iterations have been performed to finalize the inputs used during the development of the SolTrace model. These inputs include the day of the year, local solar hour, and desired number of ray intersections. Average flux and number of rays hits at the receiver surface are considered as the criterion of optimization to decide these inputs.

The receiver for the ganged heliostat system is designed in three steps (from step I to step III). In section 5.1, a horizontal flat plate is modelled as a receiver. Initially, plate size and position from the ground are optimized based on the plots of average flux and percentage of rays hits. Taking optimized plate size as a diameter of the cylinder, a cylindrical receiver is modelled in section 5.2. Now, the length of the cylinder is varied to decide the optimum length. Receiver positions in the vertical and horizontal planes are also varied to achieve better heat flux distribution. It was seen that the heat flux is concentrated at the centre and bottom surface of the cylinder. The reason behind this type of distribution was that very few rays were striking the receiver's upper surface. To overcome this problem, an inverted frustum-shaped conical receiver is designed in section 5.3. A spreadsheet program is used to find out the upper diameter (D) and semi-cone angle (α) of the inverted frustum. The inverted frustum receiver is modelled in SolTrace software using the same inputs as the cylindrical receiver and results obtained are compared with the cylindrical receiver.

4. Overview of SolTrace

SolTrace, a software developed by the National Renewable Energy Laboratory (NREL) for use in solar power based optical systems and to study and analyse their performance. This programming tool is based on the Monte Carlo ray-tracing methodology. It is available on both platforms, Windows and Mac OS [17]. Software code is based upon the ray-tracing theory of spencer et al. [18]. The results get displayed in the form of flux maps, scatter plots and statistical values. There are different phases available in the software to model the system and obtain the results as shown in figure 2. These phases are Sun, Optics, Geometry, Trace, Intersections, Flux maps and Data respectively.

Figure 2. Flowchart showing phases available in SolTrace 3.1.0 (64 bit)

The first phase in the SolTrace project is to create the sun shape and direction. There are two options available in the software to define sun position. The primary option is by defining global coordinates and the second option is to define a particular location latitude and time (the day of the year and local solar hour). There are three options available which are Gaussian, Pillbox and User-defined to create sun shape. The sets of optical properties are defined for each stage in the optics phase. In the geometry phase, the system geometry made consists of stages and elements within those stages. The optical properties which were defined at the optics phase, are linked to each element in this phase.

Once the geometry has been created, the model is traced at the trace phase. At this phase, a specific number of rays for tracing are selected by the user and every ray gets traced through the system while coming across different optical interactions. This value depends on the details needed within the results. Generally, ray numbers of the order 1 million are needed for flux mapping. The accuracy of the results is directly proportional to the number of rays being traced but the high number of rays causes an increase in the computing time. Once the system has been traced, the results are viewed in the intersection and flux maps phases. At the intersection phase, a 3-D model gets generated with stages and elements defined in the geometry phase. Generated model can be viewed with or without ray intersections. There are provisions to include the missing rays (highlighted by red colour) and to change the colour of stages and elements. At the flux maps phase, flux intensity and surface plot are shown on the right-hand side and available elements with the number of rays hits on the left-hand side. The statistical information such as peak flux (with uncertainty), minimum flux, sigma flux, average flux (with uncertainty), centroid, etc. is shown at the bottom. In the data stage, all the information generated by SolTrace, ray intersection locations, and directions at various elements, is saved to a file for reporting and/or to allow post-processing with other software [17].

5. Development of SolTrace models

5.1 Step I: Flat plate receiver

In the sun phase, the sun's position is determined by providing latitudinal coordinates of the location, day number of the year, and local solar hour. Pune, Maharashtra, India (18.5204° N, 73.8567° E) location has been chosen to conduct the experiment and input latitudinal coordinates. To decide the day number and time, various iterations have been performed and the results are plotted in the form of graphs. These iterations were performed using the same desired number of ray intersections (4 million) whereas direct normal irradiance (DNI) is different for each iteration. Initially, the average flux at the receiver surface is obtained by varying the twelfth day of each month (which is considered as midday of each month) from January to December. The plot presenting average flux and direct normal irradiance (DNI) versus each month is shown in figure 3.

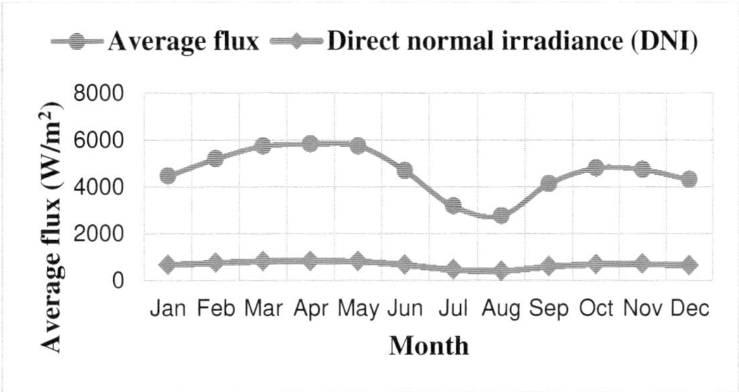

Figure 3. Average flux at receiver surface and direct normal irradiance (DNI) versus each month

From the average flux at receiver surface versus month plot, it is seen that the maximum value of average flux is obtained on the twelfth day of April at DNI = 808.39 W/m^2. Therefore, 12th April 2021 (Day Number = 102) has been considered as the day of the year. After finalizing the day of the year, the local solar hour is now varied from 7 am to 6 pm for day number 102. The average flux and DNI versus local solar hour graph is given in figure 4.

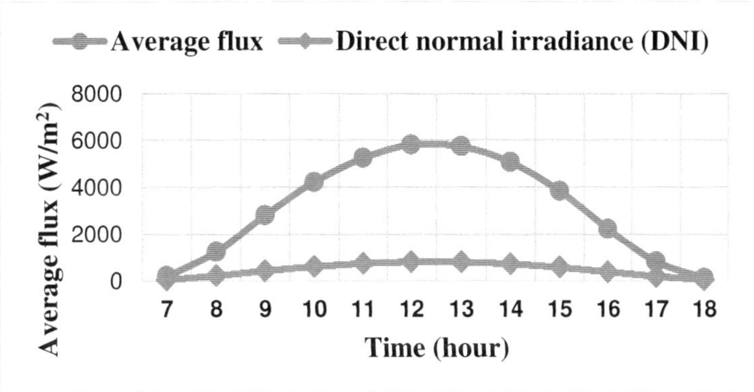

Figure 4. Average flux at receiver surface and direct normal irradiance (DNI) versus local solar hour

From the average flux at receiver surface versus local solar hour graph, it has been observed that the average flux is maximum at 12:30 pm at direct normal irradiance (DNI) = 810 W/m^2. Hence, a time equal to 12:30 pm has been considered (which is the solar noon for Pune location). The sun shape has been created by Gaussian curve with standard deviation equal to 2.73. In the next phase, optical properties of heliostat (mirror) and receiver are provided. These values are shown in Table 1.

Table 1. Optical properties of heliostat and receiver. [19]

Optical Property	Heliostat (mirror)	Receiver
Reflectivity	0.9	0.045
Transmissivity	1	1
Slope error (mrad)	0.95	0.95
Specularity error (mrad)	0.2	100

The inputs provided at sun phase (latitude = 18.53, day = 102 and hour = 12.3), optical properties mentioned in Table 1, desired number of ray intersections = 4,000,000, and DNI = 810 W/m^2 are kept the same for further iterations. Geometry has been defined on the basis of the specifications of ganged heliostat system (refer to section 2). Heliostat field and the central receiver are defined as stage I and stage II respectively. Heliostats (mirrors) are defined as elements within stage I from element = 1 to element = 16. A global origin [X Y Z] with coordinates [0 0 0] is defined and positioned on the ground at the centre of the system. The X-axis is oriented towards west, the Y-axis is oriented in the vertical direction and the Z-axis is oriented towards north [20]. The position of each mirror and central receiver is taken with reference to the global origin.

A spreadsheet program has been made to calculate the aim point positions for sixteen heliostat elements. To calculate the aim point positions, the aiming point vectors are required. The aiming point vectors describe the direction in which the sunrays will be reflected by the heliostat. The aiming point vector for each heliostat varies depending on its position and orientation in the field. The normal drawn from the heliostat is required to define them equivalent to each other in order to obtain the aiming point vector of each heliostat element. The target unit vector is calculated using the target's position relative to the heliostat whereas the heliostat normal unit vector is obtained using the solar unit vector. The equations used are generalized mathematical formulae taken from SolTrace software help file.

Initially, the distance between target and the heliostat (slant range) is obtained as,

$$L=\sqrt{(X_H-X_T)^2+(Y_H-Y_T)^2+(Z_H-Z_T)^2} \tag{1}$$

Then, the target unit vectors are calculated as,

$$X_t=\frac{X_H-X_T}{L},\quad Y_t=\frac{Y_H-Y_T}{L},\quad Z_t=\frac{Z_H-Z_T}{L} \tag{2}$$

Further, the aim point unit vector calculations are performed as,

$$X_a=\frac{(X_t-X_S)}{\sqrt{(X_H-X_S)^2+(Y_H-Y_S)^2+(Z_H-Z_S)^2}} \tag{3}$$

$$Y_a=\frac{(Y_t-Y_S)}{\sqrt{(X_H-X_S)^2+(Y_H-Y_S)^2+(Z_H-Z_S)^2}} \tag{4}$$

IOP Conf. Series: Earth and Environmental Science 1042 (2022) 012015 doi:10.1088/1755-1315/1042/1/012015

$$Z_a = \frac{(Z_t - Z_S)}{\sqrt{(X_H - X_S)^2 + (Y_H - Y_S)^2 + (Z_H - Z_S)^2}} \quad (5)$$

The aim point position is obtained in the following way,

$$X_A = X_H + 10X_a \quad (6)$$
$$Y_A = Y_H + 10Y_a \quad (7)$$
$$Z_A = Z_H + 10Z_a \quad (8)$$

The aim point positions for each heliostat are calculated and provided in the geometry phase. Mirror size (0.37 x 0.37 m) is provided in the aperture section and the nature of the surface of the mirror (i.e. flat) is given in the surface section. Each element is connected with one amongst the optical property sets that were defined earlier. After creating the heliostat stage (stage I), the central receiver stage i.e. stage II is defined. In stage II, receiver type (shape), size, and position (vertical height from the ground) needs to be determined. Various iterations were performed to optimize these parameters. The receiver type is selected as a flat plate of square shape. This shape is chosen to maintain symmetry and optical efficiency [21]. The plate size has been varied from 0.4 m to 0.6 m in the interval of 0.02 m each. Average flux at receiver surface and percentage of rays hits are plotted against the plate size as shown in figure 5.

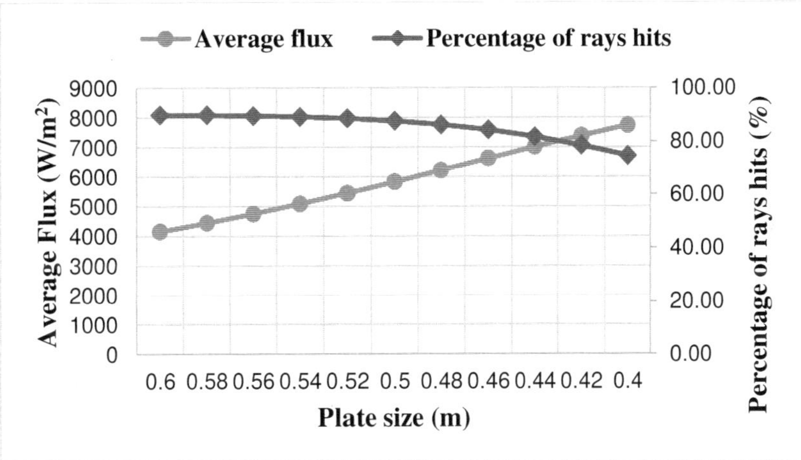

Figure 5. Average flux and percentage of rays hits versus plate size

The average flux is plotted on the primary axis as shown by the green coloured curve and the percentage of rays hits is plotted on the secondary axis highlighted by the blue coloured curve. This colour-coding has been kept constant for all iterations performed in the flat plate receiver model. It has been observed that average flux and number of rays hits are in good correlation with each other at the plate size equal to 0.43 m. Therefore, a plate side of 0.43 m has been considered for further analysis. The receiver position is finalized by varying plate height from the ground (Y-axis direction). The plate height is varied from 1.4 m to 3 m as shown in figure 6. Average flux at the plate surface and percentage of rays hits are plotted against the receiver position. From figure 6, it has been concluded that the receiver should be kept at Y = 2 m to get optimized average flux and number of rays hits. These optimized inputs are provided to stage II to complete the geometry phase.

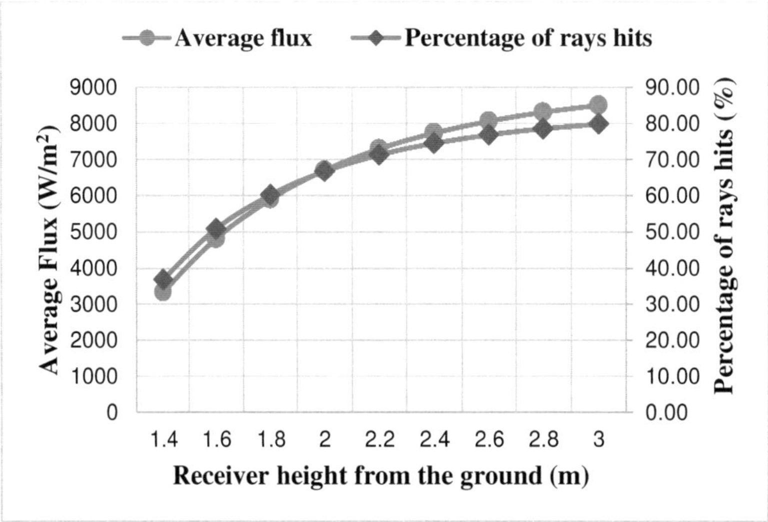

Figure 6. Average flux and percentage of rays hits versus receiver height from the ground

The desired number of ray intersections is optimized and entered in the trace phase. Average flux at the plate surface and percentage of rays hits are obtained by plotting against the total number of rays provided as observed in figure 7. The desired number of ray intersections are varied from 1 million to 5 million in the interval of 0.5 million each. It is observed that the average flux and number of rays hits gets increased up to 4 million number rays. Less variation is observed in average flux after 4 million rays. Therefore, the desired number of ray intersections is selected as 4 million.

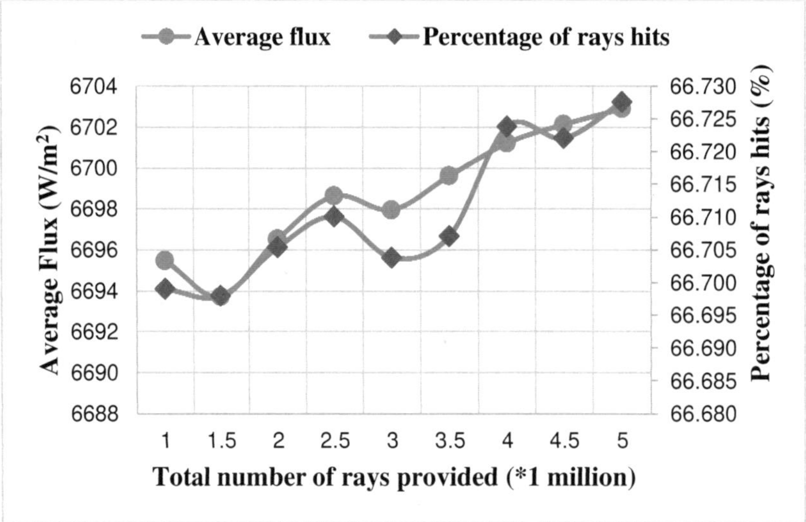

Figure 7. Average flux and percentage of rays hits versus desired number of ray intersections

After tracing the system, the model gets generated in the intersection phase. A generated model with ray intersections and the flux intensity plot are shown in figure 8 (a) and figure 8 (b) respectively.

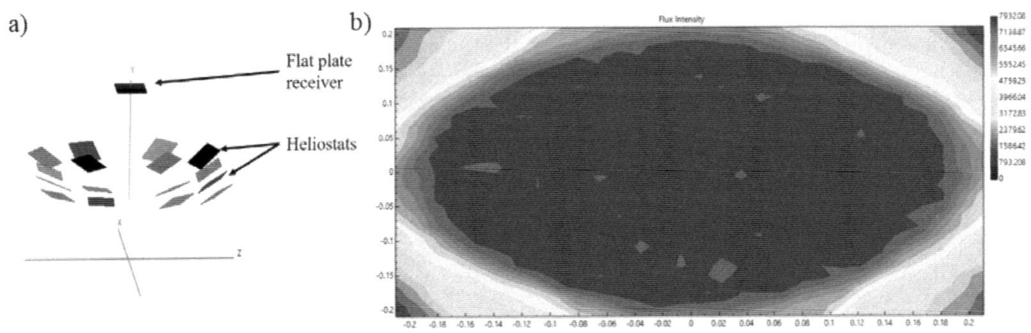

Figure 8. (a) Generated flat plate model in the intersection phase (with ray intersections) and (b) Screenshot showing flux intensity plot

The number of rays hits is 2668953 (66.72 %). Peak flux obtained is 7932.08 W/m^2 with uncertainty +/- 1.6075 % and average flux obtained is 6448.2 W/m^2 with uncertainty +/- 0.03625 %. From the result, it has been observed that average flux gets significantly increased after optimizing inputs given at different stages and receiver parameters. These inputs and receiver parameters are taken as references for modelling cylindrical receiver in the sub-section 5.2.

5.2 Step II: Cylindrical receiver

In the previous step, a SolTrace model with optimized inputs gets developed. The same inputs are used to model a cylindrical receiver in this step. Initially, the cylinder size (diameter and length) is determined. The optimized size of a flat plate is taken as the diameter of cylinder. The length of cylinder is varied from L = 0.4 m to L = 0.9 m. Average flux at receiver surface and percentage of rays hits are plotted against cylinder length as shown in figure 9.

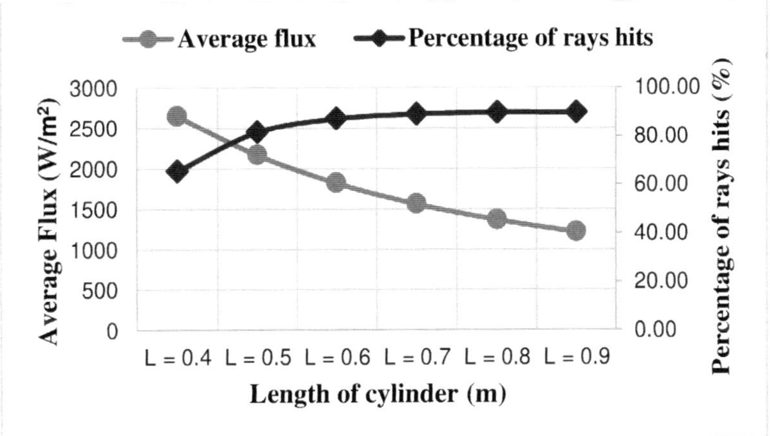

Figure 9. Average flux and percentage of rays hits versus length of the cylinder

Average flux is plotted on the primary axis shown by red coloured curve and percentage of rays hits curve is plotted on secondary axis highlighted by blue colour. This colour coding is kept constant for all iterations performed in the cylindrical receiver model. It has been observed that the average flux and the number of rays hits are in good correlation with each other at length equal to 0.5 m. Therefore, a cylinder length equal to L = 0.5 m is selected. After finalizing cylinder size, its position in a vertical direction (Y-axis) and horizontal direction (Z-axis) is determined. From the flat plate model, it has been observed that the receiver should keep at Y = 2 m (refer to figure 6) and the same has been observed for the cylindrical model. The receiver position in the horizontal direction (Z-axis) is kept at five positions called left (Z = 0), mid-left (Z = 0.105), centre (Z = 0.21), mid-right (Z = 0.345) and right (Z = 0.43) respectively. These receiver positions have been shown in figure 10. Both average heat flux and percentage of rays hits are obtained for these positions. A combination of clustered column chart and line chart is used to represent average flux and percentage of rays hits as shown in figure 11. Both the curves get increased as the receiver gets shifted towards the centre, attains a peak at the centre and then gets decreased as the receiver moves away from the centre. This iteration concludes that the receiver should be kept at the centre of the system for better flux distribution.

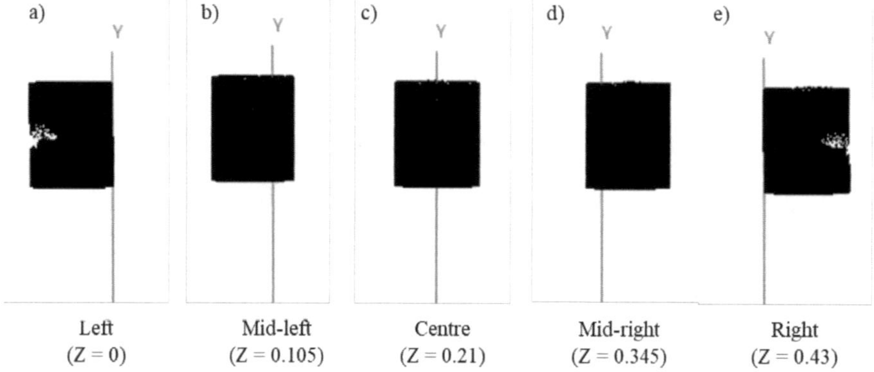

Figure 10. Cylindrical receiver position in horizontal direction (Z-axis)

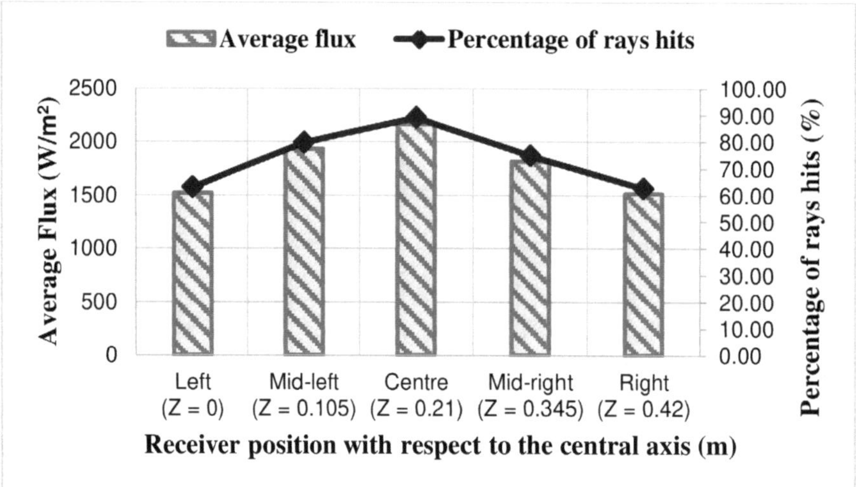

Figure 11. Average flux and percentage of rays hits versus receiver position in horizontal plane

The generated cylindrical receiver model with ray intersections has been shown in figure 12 (a) and the flux intensity plot has been shown in figure 12 (b). About 3560400 (89.01 %) of rays hit the receiver surface. The peak flux that has been obtained is 5265.57 W/m^2 with uncertainty +/- 1.0199 % and average flux that has been obtained is 2162.85 W/m^2 with uncertainty +/- 0.03622 %. OriginPro 8.0 software has been used to plot the 3D surface plot and contour plot using the flux intensity values extracted from data phase. These plots have been shown in figure 13 (a) and figure 13 (b) respectively.

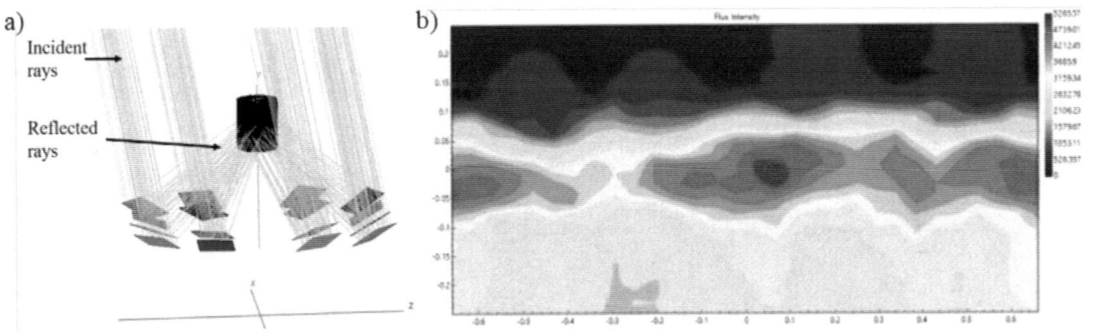

Figure 12. (a) Generated cylindrical receiver model in the intersection phase (with ray intersections) and (b) Screenshot showing flux intensity plot obtained in the flux map phase

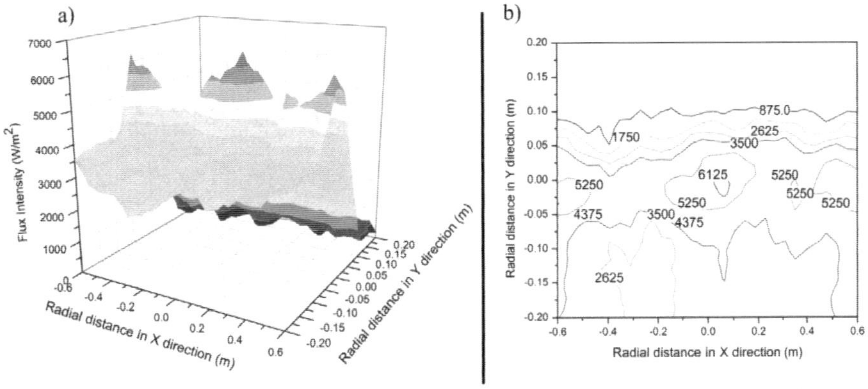

Figure 13. (a) 3D surface and (b) contour plot for the cylindrical receiver

From the flux intensity plot, it has been observed that the heat flux distribution is majorly concentrated at the centre that is a slight bit at the bottom and very little on the top of the receiver surface. The main reason behind this flux distribution is that the rays do not get hit at the top surface. To increase the number of rays hits or to accommodate most of the reflected rays, a tapered shape has been introduced in the next step of the receiver design.

5.3 Step III: Inverted frustum-shaped conical receiver

In the previous step, the cylindrical receiver has been modelled and its size and position have been iterated to optimize the average flux. Because very few rays hit the upper part of the cylinder, the heat flux variation was not uniform across the cylinder surface. In order to resolve this problem, an inverted frustum shaped conical receiver has been introduced, designed and modelled in this step. To design the inverted frustum shape, parameters such as bottom diameter, vertical height (length) and upper diameter need to be determined. These terminologies have been mentioned in figure 14 (a). The bottom diameter and vertical height (length) of the cylinder were determined in section 5.1 and section 5.2 respectively (refer to figure 5 and figure 9). The upper diameter (D) has been finalized by the iterations performed in the spreadsheet calculations as shown in Table 2. An inverted frustum receiver along with nearest (N) and farthest (F) heliostat has been drawn in AutoCAD 2019. A 2 D CAD drawing with inverted frustum receiver (sky blue line), heliostats (small circles) have been shown in figure 14 (b).

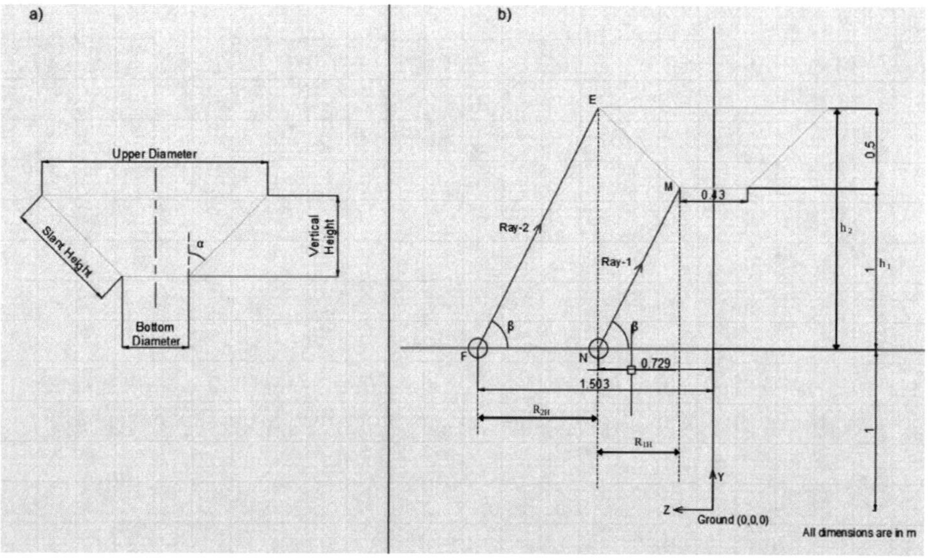

Figure 14. 2D CAD drawing showing (a) schematic representation of inverted frustum shape and (b) nearest and farthest heliostats with inverted frustum receiver.

Two rays have been plotted viz. ray1 is from the nearest heliostat to the edge of the bottom diameter and ray2 is from the farthest heliostat to the edge of the upper diameter. A ray inclination angle (β) has been defined as the angle between rays (joining heliostat to receiver surface) and horizontal. It remains the same for ray1 and ray2. The horizontal component of both the rays has been found using equation (9) and equation (11). Using a horizontal component of ray1, the bottom diameter has been found using equation (10) and the upper diameter has been calculated from the horizontal component of ray2 as provided in equation (12).

$$R_{1H} = h_1/\tan\beta \tag{9}$$

$$d = (0.729 - R_{1H}) \times 2 \tag{10}$$

$$R_{2H} = h_2/\tan\beta \tag{11}$$

$$D = (1.503 - R_{2H}) \times 2 \tag{12}$$

A pair of lower and upper diameter has been found for each inclination angle varying from 0° to 90° as shown in Table 2. The bottom diameter has been matched with the optimized bottom diameter value from the SolTrace analysis. It got a match with the value present in row number 9 (β = 63°). Hence, the pair containing d = 0.439 and D = 1.477 has been selected. The upper diameter selected is D = 1.477 m. The semi-cone angle and the slant height is found by equation (13) and equation (14).

$$\alpha = \tan^{-1}\frac{(D-d)/2}{h} \tag{13}$$

$$s = \sqrt{\left(\frac{D-d}{2}\right)^2 + h^2} \tag{14}$$

Table 2. Table showing calculations performed to find the upper diameter of the inverted frustum receiver

Sr. No.	β (°)	h_1 (m)	h_2 (m)	R_{1H} (m)	d (m)	R_{2H} (m)	D (m)
1	55	1	1.5	0.700	0.058	1.050	0.905
2	56	1	1.5	0.675	0.109	1.012	0.982
3	57	1	1.5	0.649	0.159	0.974	1.058
4	58	1	1.5	0.625	0.208	0.937	1.131
5	59	1	1.5	0.601	0.256	0.901	1.203
6	60	1	1.5	0.577	0.303	0.866	1.274
7	61	1	1.5	0.554	0.349	0.831	1.343
8	62	1	1.5	0.532	0.395	0.798	1.411
9	**63**	**1**	**1.5**	**0.510**	**0.439**	**0.764**	**1.477**
10	64	1	1.5	0.488	0.483	0.732	1.543
11	65	1	1.5	0.466	0.525	0.699	1.607

The semi-cone angle has been found equal to 24.13°. The inverted frustum receiver has been modelled with $\alpha = 24.13°$ while keeping the remaining inputs the same. The generated ray intersection model in the intersection phase has been shown in figure 15 (a) whereas flux intensity contour plot has been shown in figure 15 (b). In this figure, the vertical axis represents the height and the horizontal axis represents the curved surface (circumference) of the inverted frustum. Peak flux obtained is 3674.13 W/m^2 with uncertainty +/- 1.2515 % while average flux obtained is 2232.03 W/m^2 with uncertainty +/- 0.03624 % at direct normal irradiance (DNI) equal to 810 W/m^2. Also, about 3641200 (91.03 %) of rays get incident on the receiver surface.

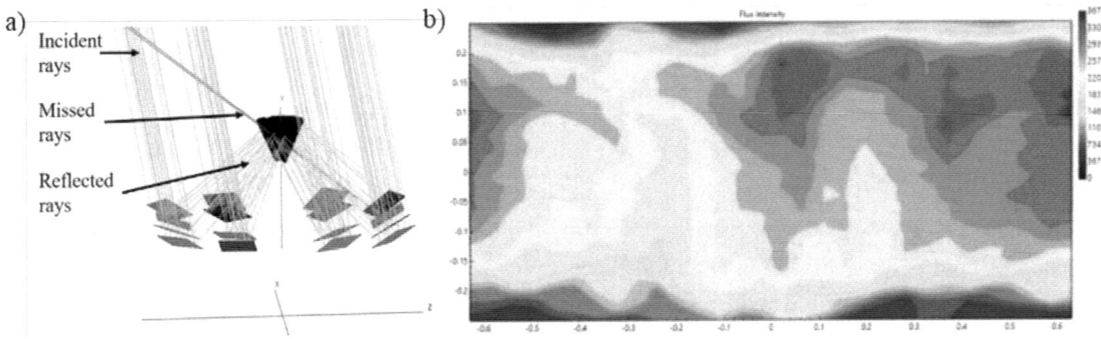

Figure 15. (a) Generated inverted frustum-shaped conical receiver model in the intersection phase (with ray intersections) and (b) flux intensity plot of the receiver

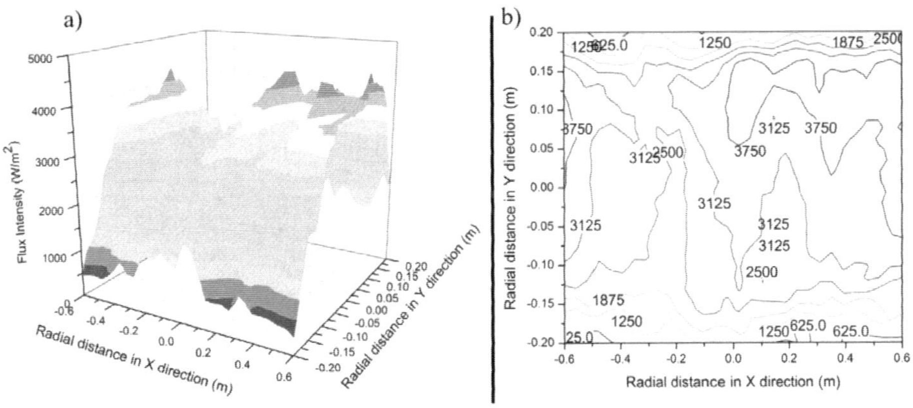

Figure 16. (a) 3D surface and (b) contour plot for inverted frustum-shaped conical receiver

A 3-D surface and contour plot of the flux intensity has been shown in figure 16 a) and figure 16 b) respectively. From different flux intensity plots, it has been seen that the heat flux distribution has been evenly spread over the curved area of the inverted frustum receiver. It has been observed that the number of rays hits at the receiver surface got increased from 89.01 % to 91.03 % when compared with the cylindrical receiver model whereas the average flux also got increased from 2162.85 W/m^2 to 2232.03 W/m^2.

6. Results

Based on the results obtained in the presented numerical analysis, the flat plate, cylindrical and inverted frustum-shaped conical receiver are compared as shown in Table 3. The desired number of ray intersections, DNI and receiver size and position are kept the same during this comparison.

Table 3. Comparison of cylindrical receiver and inverted frustum-shaped conical receiver.

Sr. No.	Parameter	Flat plate receiver	Cylindrical-shaped receiver	Inverted frustum-shaped conical receiver
1	Number of rays hits	2668953	3560400	3641200
2	Percentage of rays hits (%)	66.72	89.01	91.03
3	Peak flux (W/m^2)	7932.08	5265.57	3674.13
4	Average flux (W/m^2)	6448.2	2162.85	2232.03
5	Flux distribution	Focused at centre	Focused at centre	Even

In flat plate receiver, maximum of peak flux and average flux is obtained. Number of rays hits and their percentage has significantly increased from flat plate receiver to cylindrical receiver. Compared with cylindrical receiver, in the inverted frustum shaped-conical receiver, 80800 (2.02 %) number of rays hits gets increased. Also, the flux intensity has got increased by 69.18 W/m^2 (3.10 %). From the flux map, it has been observed that the heat flux is evenly distributed over the entire area of the inverted frustum receiver.

7. Conclusion

A step-by-step method to design a central receiver for ganged heliostat system using SolTrace 3.1.0 software has been presented. From the comparison of numerical analysis for three different types of receivers it is observed that, maximum amount of heat flux is obtained for flat plate receiver. This receiver system is quite useful for north-south heliostat field. For, spherical heliostat field which is used in central receiver system, the receiver of inverted frustum shape showed more heat flux distribution and number of rays hits at the receiver surface. Due to its tapering shape, all reflected rays get incident normally on the receiver surface which enhances the average flux. The numerical simulation by the SolTrace software is quite useful in defining and optimizing the size and position of a central receiver before its fabrication. By adopting this method in the initial phase of CRS design, a significant amount of time and capital in the development and testing of preliminary prototypes can be saved.

Acknowledgement

The author(s) received no financial support for the research, authorship, and/ or publication of this article.

References

[1] N. Jenkins and J. Ekanayake, *Renewable Energy Engineering*, 1st ed. Cambridge University Press, 2017.

[2] K. A. Soteris, *Solar Energy Engineering: Processes and Systems*, 2nd ed. Academic press, 2013.
[3] M. Geyer and W. Stine, *Power From The Sun*. 2001.
[4] Z. A. Hussaini, P. King, and C. Sansom, "Numerical simulation and design of multi-tower concentrated solar power fields," *Sustain.*, **12(6)** pp. 13–16, 2020, doi: 10.3390/su12062402.
[5] P. M. Gadhe, S. N. Sapali, and G. N. Kulkarni, "Experimental evaluation of the solar flux distribution on the flat receiver of a model heliostat system," *Int. J. Renew. Energy Res.*, **8(2)** pp. 878–887, 2018.
[6] R. P. Praveen, "Performance analysis and optimization of central receiver solar thermal power plants for utility scale power generation," *Sustain.*, 12(1, p. 16, 2020, doi: 10.3390/SU12010127.
[7] O. Behar, A. Khellaf, and K. Mohammedi, "A review of studies on central receiver solar thermal power plants," *Renew. Sustain. Energy Rev.*, **23**, pp. 12–39, 2013, doi: 10.1016/j.rser.2013.02.017.
[8] A. Soni, A. Patel, M. Pandey, and A. Gour, "Overview of different solar receiver on basis of its configuration and heat transfer fluid," *Int. J. Creat. Res. Thoughts*, 5(3, pp. 522–528, 2017.
[9] L. Chen *et al.*, "Thermal analysis of a volumetric solar receiver," *J. Therm. Sci.*, **28(6)** pp. 1176–1185, 2019, doi: 10.1007/s11630-019-1220-7.
[10] L. Prattico, S. Amicabile, M. Roccabruna, L. Crema, and F. Alberti, "Optimization study of volumetric receivers for high temperature solar plants," *30th Int. Conf. Effic. Cost, Optim. Simul. Environ. Impact Energy Syst. ECOS 2017*, no. December, p. 16, 2017.
[11] S. B. Bopche and S. Kumar, "Experimental investigations on thermal performance characteristics of a solar cavity receiver," *Int. J. Energy Environ. Eng.*, **10(4)** pp. 463–481, 2019, doi: 10.1007/s40095-019-00321-4.
[12] P. M. Gadhe, A. G. Ghoti, S. N. Sapali, and G. N. Kulkarni, "Experimental method to find the flux and temperature distribution on the flat receiver of small central receiver system," *Int. Rev. Mech. Eng.*, **11(6)** pp. 426–432, 2017, doi: 10.15866/ireme.v11i6.12880.
[13] T. Tan and Y. Chen, "Review of study on solid particle solar receivers," *Renew. Sustain. Energy Rev.*, 14(1, pp. 265–276, 2010, doi: 10.1016/j.rser.2009.05.012.
[14] H. Shatnawi, C. W. Lim, and F. B. Ismail, "Solar tower power: the impact of external receiver on optimal performance and energy storage," *Int. J. Eng. Adv. Technol.*, **9(2)** pp. 4195–4202, 2019, doi: 10.35940/ijeat.b4929.129219.
[15] T. Wendelin, "Soltrace: A new optical modeling tool for concentrating solar optics," *Int. Sol. Energy Conf.*(March 2003, pp. 253–260, 2003, doi: 10.1115/ISEC2003-44090.
[16] R. Patwardhan and R. Pandit, "Solar central receiver system employing common positioning mechanism for heliostats," US 9568215 B2, 2017.
[17] T. Wendelin, A. Dobos, and A. Lewandowski, "SolTrace: A ray-tracing code for complex solar optical systems," United states, 2013. [Online]. Available: www.nrel.gov/publications.
[18] E. C. Doerner, "General ray-tracing procedure," in *Journal of the Optical Society of America*, **52**, no. 1962, United states: OSA Publishing, 1961, pp. 672–678.
[19] P. Gadhe, S. Sapali, and G. Kulkarni, "Experimental and numerical investigations of solar flux density distribution over flat plate receiver of model heliostat system," *J. Therm. Eng.*, **6(6)** pp. 312–322, Nov. 2020, doi: 10.18186/thermal.831343.
[20] M. Maliage and T. H. Roos, "The flux distribution from a 1.25 m^2 target aligned heliostat: comparison of ray tracing and experimental results," *Proc. SASEC*, no. September 2016, pp. 1–9, 2012.
[21] L. Li, J. Coventry, R. Bader, J. Pye, and W. Lipiński, "Optics of solar central receiver systems: a review," *Opt. Express*, **24(14)** p. A985, 2016, doi: 10.1364/oe.24.00a985.

IOP Conf. Series: Earth and Environmental Science 1042 (2022) 012016 doi:10.1088/1755-1315/1042/1/012016

Techno-Economic Assessment of Grid Connected Solar PV/Wind hybrid System

Mustapha Sulley[1], Amevi Acakpovi[2], Patrick Adjei[1], David M. Sackey[3], Felix Offei[1], Maxwell Afonope[1], Daniel Kofi[1], Godfred Tay[1]

[1] Regional Maritime University, Department of Electrical and Electronic Engineering, Ghana
[2] Accra Technical University, Department of Electrical and Electronic Engineering, Accra, Ghana
[3] Ho Technical University, Department of Electrical and Electronic Engineering, Ho, Ghana

*Email: patrickadjei1@gmail.com

Abstract. Wind and Solar PV energy as alternative energy supplies to the traditional fossil fuel have been a subject of study for researchers at various fora including climate change summits. However, the technical and economic feasibility of wind and solar projects involve a lot of complexity and depends mostly on geographical location and availability of resources. To address the constraints and factors affecting wind and solar PV systems, this paper aims to undertake a techno-economic feasibility assessment of a grid connected solar PV/Wind hybrid system capable of meeting a typical commercial load, located in Kumasi, Ghana. To achieve the purpose of this study an energy audit has been undertaken to establish the load demand of the facility. Again, RETSCREEN software was used to design and simulate the proposed hybrid system in order to analyse the technical, economic and environmental implications of the system. Findings show that, there is a high potential for providing commercial scale energy consumption in most months in the year, and there is extra energy available to be sold to the grid, generating considerable income. Though the system has a high levelized cost of energy (LCOE) as compared to the existing grid tariff, the study adds significantly to the national objective to reduce its dependency on fossil fuels while meeting local energy requirement.

1. Introduction

Important, worldwide debates have concentrated on problems such as sustainable energy development and greenhouse gas (GHG) emissions in recent years. The International Energy Agency (IEA) has stated that, assuming current government policies remain unchanged, the world's primary energy demand is predicted to rise by 1.5 percent each year from 2007 to 2030, for a total increase of roughly 40 percent [1]. Global electricity demand, on the other hand, is predicted to rise by 2.5 percent every year over the same time period [2] However, fossil fuels continue to dominate the world's fuel mix, resulting in continuous increases in carbon dioxide in the atmosphere, exacerbating the negative effects of global warming.

Conventional energy sources, such as fossil fuels, are rapidly decreasing, requiring international leaders to implement more sustainable energy policies to control this trend [3]. Renewable energy has been considered as one of the primary solutions to the world's energy crisis [4]. Renewable energy sources are often carbon-free or carbon-neutral, making them one of the most effective ways to help alleviate the consequences of climate change. The biggest disadvantage of renewable energy technology is the expensive initial cash required [5]. The costs of these renewable energy sources, on the other hand, have been declining in recent years as a result of technological advancement and market maturity [5].

Most economies benefited greatly from this cost decrease from the use of those technologies were encouraged. This eventually led to sustainable energy development and the mitigation of the effects of climate change. Global renewable energy capacity rose at rates ranging from 10% to 60% per year for many technologies throughout the five-year period spanning from the end of 2004 to the close of 2009 [6]. However, grid-connected solar photovoltaic (PV) and wind systems continue to be the world's fastest growing energy technologies, with cumulative installed capacity increasing by 60% per year from the end of 2004 to the end of 2009 [7].

In the past few years, the people of Ghana had to endure some intermittent power outages and this was partially due to over dependence on hydropower which suffers from low water volume during the dry season. Again, the insufficient supply of fuel to thermal power plants and foreign exchange challenges are partially responsible for these intermittent power outages. Grid power demand is expected to increase beyond 24,000GWh by 2023, from around 6,900GWh in 2000 [8]. To ensure reliable, uninterrupted energy supply by 2030, Ghana's present capacity, which was 1760 MW in 2006 must be increased to meet growing energy demand by 2030 [8]. Ghana aims towards increasing the percentage of alternative clean and renewables in it energy mix to 10% by 2030 [8].

A glance at the global mean solar radiation reveals that, the continent of Africa experiences the most intense sun radiation, which ranges from 300 to 350 W/m2 annually, and moderation to commercial scale wind speeds, making a combination of solar PV and wind power system a great option for hybrid energy supply [9,10]. Ghana has great potentials for solar energy such as standalone and grid connected solar PV system. Despite these enormous potentials, Ghana has not made enough use of its renewable energy potentials and still suffer from intermittent power outages while many communities do not still have access to grid electricity.

A plethora of studies has been conducted in the field of grid connected hybrid renewable energy systems all around the world as elaborated in the paragraphs below.

At Hashemite University, [11] examined the economic and technical performance of an open-loop and double axis fixed solar PV system in Jordan. The grid connected solar PV system was technically evaluated based on the final energy yield and conversion efficiency. The economic evaluation considered for the grid connected solar PV system were internal rate return, payback period and cost of electricity per kilowatt hour. The result of the study revealed that the solar PV with the tracker performed better in energy yield than the fixed axis solar PV system.

Pradhan et al. (2015) evaluated a grid connected solar PV system for a Hostel and an engineering college in India [12]. The authors used RETScreen computational software to simulate the design. The study objective was to examine the amount of reduction in energy consumption from the grid system with the introduction of solar PV power into the energy mix. According to the study, 772 gallons of crude oil was saved from the adoption of the grid connected solar PV system. However, the study could not establish the base emission of the study area.

Another study was conducted by [13] to see how much electricity may be saved in Jordan's water industry by using grid-connected PV systems. The study was held in five different locations, and it proved that grid connected PV power system is a feasible and cost-effective alternative.

Kumi (2012) assessed the economic and technical impact of a grid connected solar PV system for Kwame Nkrumah University of Science and Technology (KNUST) in Kumasi, Ghana [14]. The proposed system is only viable if the LCOE is more than 0.43$/kWh according to the findings of the study. The study also discovered that the most expensive solar PV module was the amorphous silicon panels, but performed better in energy yield as compared with monocrystalline and polycrystalline modules.

Al-Masri et al. (2019) conducted a technical and economic investigation of a grid-tied wind power system in the desert of Jordan [15]. The study performed a mathematical modelling in MATLAB using hourly wind speed data and load demand of the case study area. Simulated annealing and genetic algorithm were used by the authors to find the lowest LCOE of the proposed system. The study resulted in the reduction of CO2 emission and LCOE by 18.74% and 58.39% respectively.

Similarly, Dalabee (2017) picked five Jordanian locations to assess the economic potential of wind energy plants [16]. Their best site's LCOE was 0.0259 to 0.0498 $/kWh, while the LCOEs of other sites were 0.2220 to 0.2220 $/kWh. The cost range was considered appropriate within the limitations of the MENA region's needs.

Ammari (2015) used recorded averages and the Weibull function in similar research for five locations in Jordan to measure wind power [17]. According to the authors, the greatest wind speed at Aqaba airport was 7.05 m/s, while the minimum wind speed at Azraq was 1.15 m/s. The Aqaba airport produced the most energy, with the Weibull function having shape parameter within the range of 1.65 to 2.78.

Other literatures were conducted in a combination of renewable energy systems mainly on hybrid systems which focus on combinations that include solar and wind in the energy mix.

Acakpovi et al.(2020) designed and compared two hybrid systems using HOMER software based on the same load profiles [18]. The wind energy system using hydrogen fuel cell storage had the lowest NPC compared with the simple wind system using battery as a means of storage. Again, the wind energy and hydrogen fuel cell system had the lowest LCOE compared with the wind energy system with battery energy storage system. Technically, the wind energy system with battery had more energy reserves than the wind energy system with hydrogen fuel cell.

Al-Ghussain (2018) conducted an economic and technical assessment of different mixtures of Solar PV, pumped hydro-storage, and hydrogen fuel-cell system for a university in Northern Cyprus [19]. The objective function of the optimization model for the proposed system was to have an LCOE equal the national electricity price whiles maximizing the renewable fraction of the system. The optimal system consisted of a combination of pumped hydro system of 4.14 MWh, 1.16 MWh hydrogen fuel-cell, and 2.57 MW solar PV system with an LCOE of 0.181$/kWh.

Again, Acakpovi et al. (2021) in [20], used the Anloga wind regime in Ghana to conduct an economic and technical assessment of wind and hydrogen fuel cell system. The study adopted HOMER software to design, analyze and optimize the proposed system. Technically, the proposed system was unable to meet the load during the whole by 10%. However, the system generated an annual electricity of 25,999kW for the first year of the project with an LCOE of 0.602$/kWh. The proposed system had a reduced emission compared with a diesel generator used as back-up.

Alternatively, Halawani et al. (2015) in [21], explored the viability of hybrid renewable energy systems in off-grid and remote areas with a smart microgrid capability. The developed hybrid system consisted of solar PV, diesel generator, Wind, hydrogen fuel cell, and battery. The proposed system was designed and simulated in HOMER and MATLAB. The system incorporated a control and synchronization strategy for the various configurations.

Okonkwo et al. (2017) designed and simulated an off-grid HES made of Solar PV/Wind battery storage system to satisfy the energy demand of a hotel in Jordan [22]. The proposed system was technically and economically assesses based on ability to serve the load requirement, NPC, payback period and savings-to-investment ratio. The off-grid hybrid system was technically and economically feasible as compared to the on-grid system.

Again, Vick and Neal (2012) evaluated a grid connected solar PV and wind system for water pumping system [23]. The authors simulated and analysed each RET that has the capacity to serve the load requirement. The study concluded that a standalone single RET was not as efficient in pumping water as compared to hybridized system of multiple RET. The hybridized RET of Wind and Solar PV power generation pumped about 28% more water than the single RETs.

Alharthi et al. (2018) performed a technical and economic assessment of a grid connected hybrid solar PV/wind system using HOMER software in Saudi Arabia [24]. The proposed system was designed in HOMER to meet a peak load of 239kWp and an energy demand of 15000 kWh/day. The technical parameters used to evaluate the proposed system were annual electricity production (AEP). The economic parameters of the proposed grid connected RET were the net present cost and the levelized cost of electricity. The proposed system was technically and economically feasible according to the study.

Budes et al. (2020) used HOMER Pro to conduct smilar assessments of the impact of an ON grid wind and solar PV system [25]. The HOMER software was used by the authors to access the weather data of the study area (Puerto Bolivar in Colombia) and to comprehensively design a HES system that can supply a fraction of the energy requirement of the study area. The optimal hybrid renewable configuration to combine with the thermoelectric grid power generation was a 441 Solar PV panels and 3 wind turbines (2.5kW capacity each). The economic parameters used to evaluate the proposed system was NPC with a cost of $11.8 million and a reduction in CO2 emission of 244.1tons annually.

Mahesh and Sandhu. (2020) used genetic algorithm and developed an energy filter method to optimize a hybrid power supply with battery storage system using solar PV and Wind [26]. The study aimed at reducing the total cost of the proposed system whiles ensuring a reliable power supply from the hybrid RET. The reliability index used to assess the proposed algorithm and energy filter were battery state of charge (SOC), full use of energy resources and quality of power injected into the grid. These were modelled as reliability constraints and used to size each component of the proposed system. The authors validated the proposed energy filter method to evaluate it effectiveness in sizing a hybrid RET.

Hybrid energy systems have also grown in popularity as a result of their ability to regulate resource consumption while allowing for the optimal mix of technologies to accomplish the desired outcomes [27].

The reviewed literature demonstrates the numerous researches into finding the best configuration and sizing of hybrid RETs. The economic and technical evaluation of these hybrid RETs are a major objective in analyzing the feasibility of hybrid renewable energy systems. In this regard, this paper aims at evaluating the economic and technical feasibility of a grid connected solar PV/Wind energy supply for a commercial office in Kumasi, Ghana. Though many studies have used HOMER software for these assessments, few literatures used the RETScreen that presents several and proven benefits in terms of technical, economic and emission estimation.

To the authors' best of knowledge, no similar study on hybrid solar/wind supply has been undertaken in the proposed location in Ghana before and especially with the use of RETScreen software. Indeed, an economic and technical study of a grid connected hybrid renewable energy system at Kumasi Metropolitan Assembly (KMA) will set a precedence for the development and integration of Hybrid RE system into the national grid. Again, the study will set the pace for other investors to emulate and, provide support to the national agenda of increasing the renewable energy integration to the national energy mix by 10%.

In this paper, a grid connected solar PV/wind system is designed to serve the power requirement of KMA; the design has been assessed using established techno-economic metrics. Mainly, the significance of this study relates to the location specific renewable energy resource; the specific hybrid configuration suitable for the selected area and the unique approach by using RETScreen software.

2. Materials and Methods

The renewable power capacity in the targeted city of Kumasi is shown in the following subsections, with specifications for PV modules and wind turbines, as well as system economic metrics and corresponding national grid statistics.

2.1. *Electricity Requirement of KMA Office*

The seasonal electricity consumption for KMA office is depicted in Figure 1. This was determined through walk-through energy audit and energy consumptions pattern from their utility bills. The highest peak demand in February and March were 439 kW and 455 kW, respectively. With a high demand of 330 kW, November was the month with the lowest demand. These values, on the other hand, show the maximum and minimum demands reported on a given month's day. As a result, the study considered the average annual energy use, which was roughly comparable throughout all months. Following that, a monthly average of 3,850.00 kWh was given as the daily mean energy use.

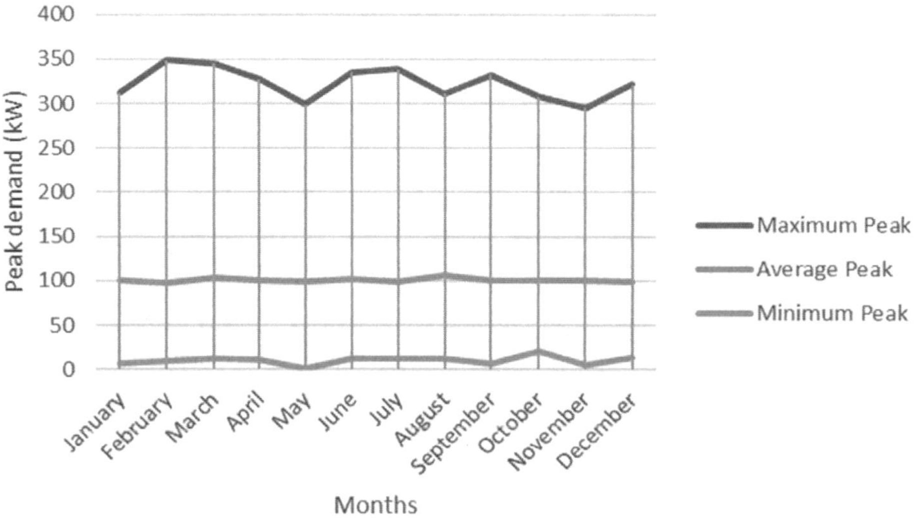

Figure 1. Scaled monthly energy demand of KMA office

2.2. *Solar and Wind Potential of Kumasi*

Kumasi's geographical position provides a tremendous chance for plentiful renewable energy resources. Kumasi central in particular, receives an average of 4.96 kWh/m^2/day of solar irradiation, with July and August are receiving the least (4.225 kWh/m^2/day) and February receives the most (5.582 kWh/m^2/day). The average wind speed and irradiance for the Kumasi central where KMA office is located are shown in Figure 2 and 3 respectively from the National Renewable energy laboratory (NREL) weather database. The average wind speeds in the Kumasi central area is 4.15m/s at 10m anemometer height with July and August recording highest wind speeds of 4.95 and 5.01m/s respectively. The lowest wind speeds were recorded in November and December with 3.4 and 3.36m/s respectively.

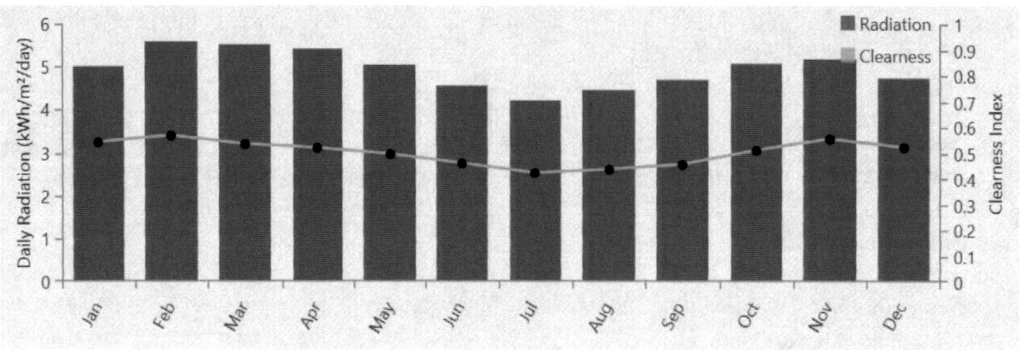

Figure 2. Solar radiation and clearness index of Kumasi central

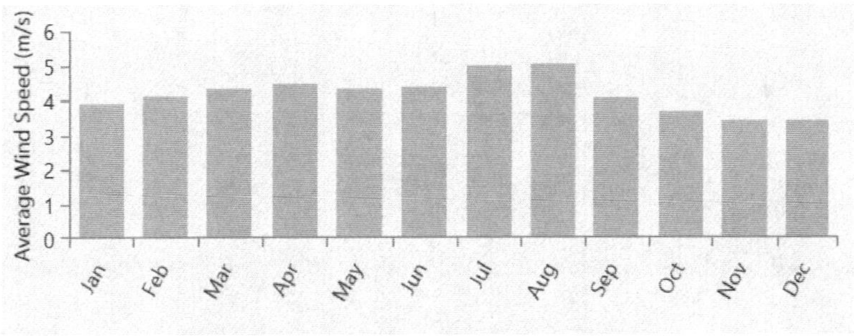

Figure 3. Wind speed of Kumasi central

2.3. Solar PV Module and Wind Turbine Details

For the PV farm, the study adopted 1000 units of polycrystalline silicon Canadian Solar, poly-Si - CS6X-300Wp module from RETSCREEN library. Each unit is rated at 300 Wp, with a total installed capacity of roughly 300 kW. Under typical test circumstances, the module efficiency is 15.63%, and it is installed in a rack-mount configuration. In this paper, the selected winf turbine has 500 kW power rating and a height of 70m. The technical specifications of the solar PV and wind turbine used in this study are provided in Figure 4 and 5. The overall configuration of the hybrid system is also shown in Figure 6.

Wind turbine

Power capacity per turbine	kW	500
Manufacturer		DeWind
Model		DEWIND 41 - 70m
Number of turbines		1
Power capacity	kW	500
Hub height	m	70
Rotor diameter per turbine	m	41
Swept area per turbine	m²	1,320.25
Energy curve data		Standard
Shape factor		2

Figure 4. Wind turbine specifications

Photovoltaic

Type		poly-Si
Power capacity	kW	300
Manufacturer		Canadian Solar
Model		poly-Si - CS6X-300P - MaxPower
Number of units		1,000
Efficiency	%	15.63%
Nominal operating cell temperature	°C	45
Temperature coefficient	% / °C	0.4%
Solar collector area	m²	1,919

Figure 5. Solar PV specifications

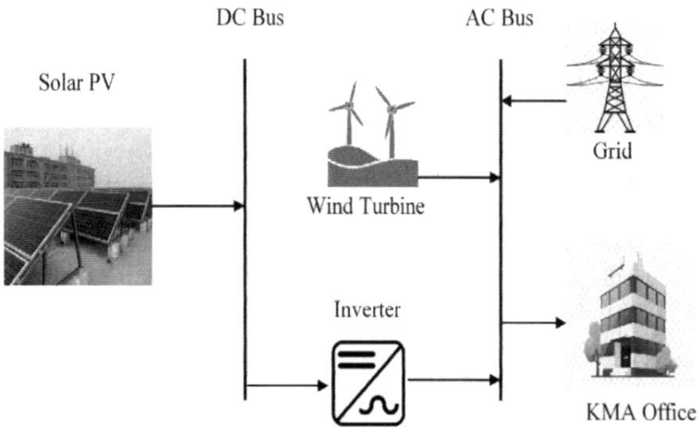

Figure 6. Schematic diagram of the proposed system

2.4. Proposed Hybrid System Economic and Electric Grid Input Parameters

The cost of the proposed grid connected hybrid solar PV and wind system is made of feasibility studies, development cost, engineering cost, balance of system cost which make up the total initial cost of the project. The annual cost which comprises the operation and maintenance cost is also presented in the summary (see Figure 7 and 8). The project is assumed to receive a government grant of US$30,000.00 according to the energy commission of Ghana. This grant is factored into the study as annual savings.

Annual costs (credits)	Unit	Quantity	Unit cost	Amount
O&M				
Show data				$ 62,600
Land lease & resource rental	project	12	$ 200	$ 2,400
Property taxes	project	1	$ 5,000	$ 5,000
Insurance premium	project	1	$ 7,000	$ 7,000
Parts & labour	project	1	$ 12,500	$ 12,500
GHG monitoring & verification	project	1	$ 2,300	$ 2,300
Community benefits	project	1	$ 3,000	$ 3,000
General & administrative	%	5.0%	$ 94,800	$ 4,740
User-defined	cost			$ -
Contingencies	%		$ 99,540	$ -
Subtotal:				**$ 99,540**
Annual savings	**Unit**	**Quantity**	**Unit cost**	**Amount**
User-defined-Subsidy	cost			$ -
Government subsidy	cost	1	$ 30,000	$ 30,000
Subtotal:				**$ 30,000**

Figure 7. Annual cost of the system

RETScreen - Cost Analysis

Initial costs (credits)	Unit	Quantity	Unit cost	Amount	Relative costs
Feasibility study					
Site investigation	p-d	1	$ 3,000	$ 3,000	
Resource assessment	project	1	$ 2,500	$ 2,500	
Environmental assessment	p-d	1	$ 3,000	$ 3,000	
Preliminary design	p-d	1	$ 4,500	$ 4,500	
Detailed cost estimate	p-d	1	$ 3,000	$ 3,000	
GHG baseline study & MP	project	1	$ 1,500	$ 1,500	
Report preparation	p-d	1	$ 2,000	$ 2,000	
Project management	p-d	1	$ 6,000	$ 6,000	
Travel & accommodation	p-trip	5	$ 2,500	$ 12,500	
User-defined	cost			$ -	
Subtotal:				$ **38,000**	0.8%
Development					
Contract negotiations	project	1	$ 1,500	$ 1,500	
Permits & approvals	project	1	$ 5,500	$ 5,500	
Site survey & land rights	project	1	$ 4,300	$ 4,300	
GHG validation & registration	project	1	$ 1,500	$ 1,500	
Project financing	project	1	$ 3,800	$ 3,800	
Legal & accounting	project	1	$ 4,500	$ 4,500	
Project management	project	1	$ 2,100	$ 2,100	
Travel & accommodation	p-trip	5	$ 2,500	$ 12,500	
Engineering					
Site & building design	project	1	$ 3,000	$ 3,000	
Mechanical design	project	1	$ 3,500	$ 3,500	
Electrical design	project	1	$ 1,500	$ 1,500	
Civil design	project	1	$ 2,500	$ 2,500	
Tenders & contracting	project	1	$ 3,000	$ 3,000	
Construction supervision	project	10	$ 500	$ 5,000	
User-defined	cost			$ -	
Subtotal:				$ **18,500**	0.4%
Power system					
Photovoltaic	kW	300	$ 3,300	$ 990,000	
Wind turbine	kW	500	$ 7,500	$ 3,750,000	
Road construction	km	3	$ 5,000	$ 15,000	
Transmission line	km	3	$ 2,000	$ 6,000	
Substation	project	1	$ 7,000	$ 7,000	
Energy efficiency measures	project	1	$ 25,000	$ 25,000	
User-defined	cost			$ -	
Subtotal:				$ **4,793,000**	98.1%
Balance of system & miscellaneous		Specific project costs			
Building & yard construction	m^2			$ -	
Spare parts	%			$ -	
Transportation	project			$ -	
Training & commissioning	p-d			$ -	
Total initial costs				$ **4,885,200**	100.0%

Figure 8. Initial cost of the system

In addition, the economic parameters of the national grid are essential for evaluating and comparing the efficacy of the proposed hybrid system. According to the Ghana Electricity Company, the average energy price for power purchased from the national grid is about 0.50 $/kWh.

2.5. Economic Assessment of the HES

The economic feasibility of any renewable project is dependent of Internal Rate of Return (IRR), Net Present Value (NPV), Payback period (PBP) and the cost of energy per kilowatt hour. These economic parameters have also been used by researchers to evaluate the economic performance of projects especially renewable energy projects. As a result, the study presents the important theoretical formulations of the criteria mentioned in the following subsections in order to conduct the economic evaluation.

2.5.1 Levelized Cost of Electricity (LCOE). Cost of energy per kilowatt-hour as stated in the above subsection is employed to assess the economic performance of the renewable energy system proposed in this study. The cost of electricity is estimated using equation (1)

$$LCOE(\$/kWh) = \frac{Total\ life\ cycle\ \cos t(\$/W_p)}{Total\ Lifetime\ yield(kWh/W_p)} \tag{1}$$

Where $LCOE(\$/kWh)$ represent the cost of electricity per kWh

kWh is Kilowatt-hour of electricity

W_p is watt-peak of solar radiation

2.5.2 *Estimation of NPV and IRR.* The NPV is an important metric in assessing HES. It refers to the current net value fo the project while considering all future earnings. The NPV is calculated as shown in equation (2)

$$NPV = \sum_{t=0}^{N} \frac{c_n}{(1+r)^n} \tag{2}$$

C represents the capital cost

n is the project lifespan

r is the interest rate

On the other hand, IRR is a frequently used metric for assessing the competitiveness of energy generation projects. Technically it is considered as the discount rate when the NPV is null and can be calculated with equation (3) as follows.

$$IRR = \sum_{t=1}^{1} \frac{R_t}{(1+r)^t} - c \tag{3}$$

Where R_t is the Net cash inflow during period t

r is discounted rate

2.5.3 *Payback Period (PBP).* The PBP is an economic parameter used to determine the payment period of the any project. In this study it is adopted to evaluate the economic viability of the project, a shorter PBP is preferred. Equation (4) is adopted to estimate the PBP of the RE project under study.

$$PBP = \frac{C}{R_1} \tag{4}$$

Where the PBP denotes payback period for the investment and R_1 is the returns from the proposed hybrid system project in terms of \$/kWh in the first year.

IOP Conf. Series: Earth and Environmental Science 1042 (2022) 012016 doi:10.1088/1755-1315/1042/1/012016

3. Results and Interpretations

The results illustrate the efficacy of the proposed system with regard to the fulfillment of the monthly energy requirement of the KMA office while also delivering a large economic benefit when compared to the price of electricity delivered by the national grid.

3.1. Electricity Production from the Proposed Syste

The annual energy output of the PV generator and wind turbine is 388MWh and 374 MWh, respectively. Figure 9 depicts the daily energy production of PV, wind systems and grid purchase. Figure 10 presents a summary of electricity production from each of the systems in a year.

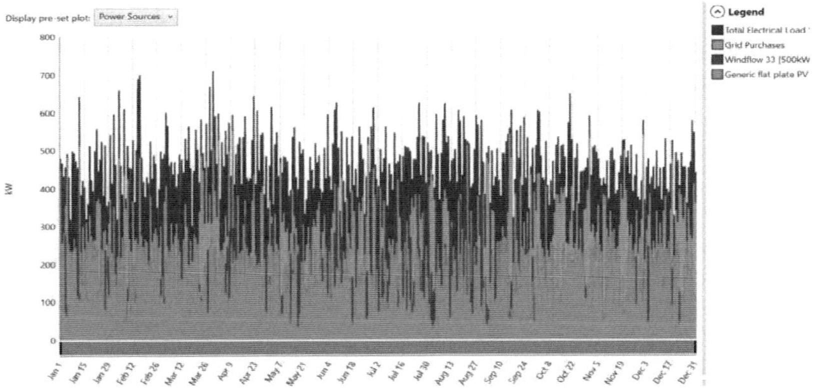

Figure 9. daily electricity production from the grid connected hybrid system

Power system - Total		
Capacity	800	kW
Electricity	762	MWh
Photovoltaic		
Capacity	300	kW
Electricity	388	MWh
Wind turbine		
Capacity	500	kW
Electricity	374	MWh

Figure 10. Electricity production and capacity of the hybrid system

3.2. System Economics

Figure 11 illustrates the proposed hybrid system's yearly cash flow. As can be seen from the graph, the simple payback for this proposed system before tax is 15 years and 6 months. Considerung the payback period, the cash flow of the proposed system is positive, with an indicated positive revenue. Figure 12 presents the cumulative cash flow of hybrid supply. The equity payback period of the project happens in the 26th year on the 8th month. The financial viability of the proposed system is presented in Figure 13. The Figure presents the IRR and PBP.

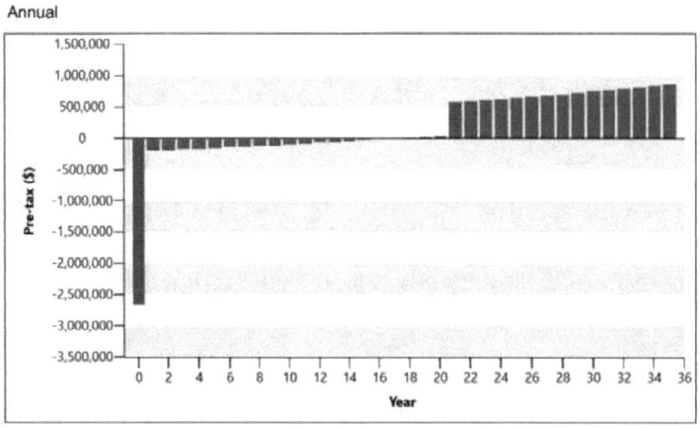

Figure 11. Annual cash flow

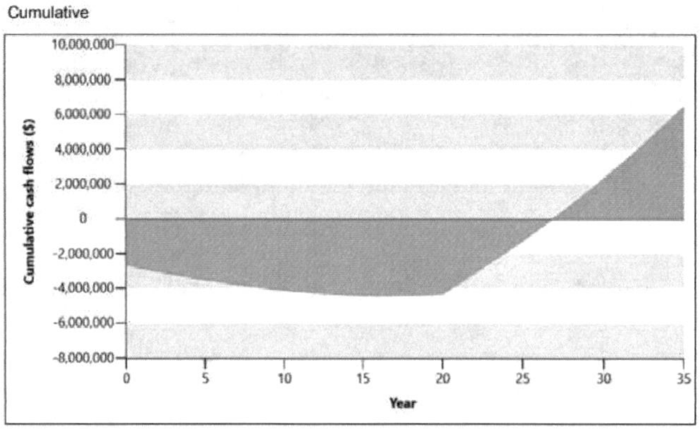

Figure 12. Cumulative cash flow

Financial viability		
Pre-tax IRR - equity	%	7.7%
Pre-tax IRR - assets	%	2.9%
Simple payback	yr	15.7
Equity payback	yr	18.6
Net Present Value (NPV)	$	-397,403
Annual life cycle savings	$/yr	-37,609
Benefit-Cost (B-C) ratio		0.73
Debt service coverage		0.85
GHG reduction cost	$/tCO$_2$	51.69
Energy production cost	$/kWh	0.671

Figure 13. Financial viability of the proposed system.

3.3. Emission Reductions

Figure 14 shows the total emission reduction of the hybrid system. The Figure show the initial base GHG emission and proposed case emission. Again, the Figure presents the GHG reduction revenue if the GHG reduction is converted into emission credit. The base case GHG emission is 176.5 tCO_2 whiles the proposed hybrid system is 12.4 tCO_2. The total reduction in GHG emission if the proposed hybrid system is implemented is 164.2 tCO_2. In simple terms, this equivalent to 381.8 barrels of crude oil not consumed.

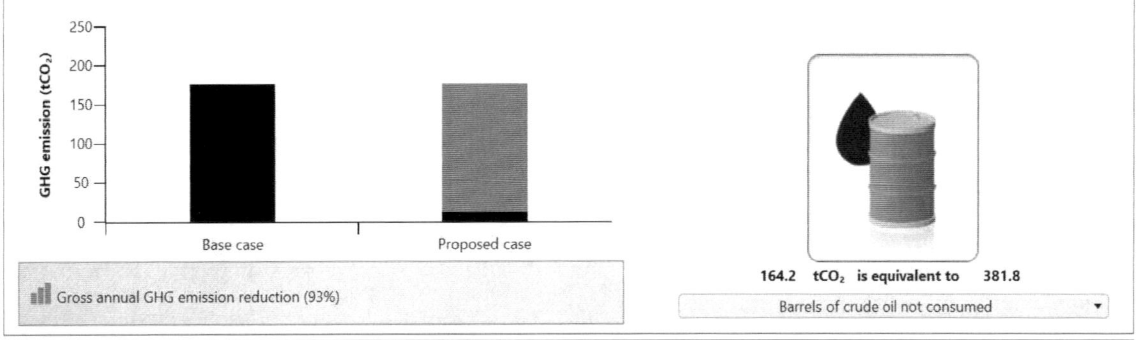

GHG reduction revenue		
GHG reduction credit rate	$/tCO₂	25.00
GHG reduction credit duration	yr	25
GHG reduction credit escalation rate	%	0.1%
GHG credits transaction fee	%	0.2%
Net annual GHG emission reduction	tCO₂	163.9
GHG reduction revenue	$	4,096

Figure 14. Emission savings

4. Discussion

The total electricity production from the proposed system shows that a single renewable energy technology could not meet the electricity demand of the KMA office effectively and throughout the year. Technically, the proposed system is more effective to supply the load as compared to a single renewable energy technology. Abujubbeh et al.(2019) performed an economic and technical analysis of a grid connected solar PV and wind system for a town in Jordan [28]. It was observed that a single renewable energy technology was not capable of suppling the load conveniently due to intermittency challenges.

Renewable energy systems in general are technically not too complex to implement, however, their economic feasibility is challenged with numerous primary roadblocks. For instance, the price of solar PVs has been considerably higher over the year, even though it has known recently a significant decrease. The technology of wind energy conversion has been limited to well developed countries, making them quite expensive in developing countries with these resources available. The technology to convert heat of the sun (CSP) for power generation is not well established in Ghana just as tidal and wave energy technologies. These economic roadblocks make RET projects, excessively expensive in terms of profitability and shorter project life. In countries where feed-in-tariff exists, the feed-in-tariff per kWh of electricity are not investment friendly.

In this paper, a PV and wind hybrid system with PBP of 15.6 years and LCOE of 0.671 $/kWh was proposed, designed and simulated. The system presented greater benefits than the figures obtained in the study of [28] and [22] though the economic parameters seem marginally different. This paper proposes a feasible solar PV and wind power system that can be coupled to grid to support existing power generation capacity. Furthermore, the Net Present Value (NPV) is US$ -397403 with an internal rate of return (IRR) of 7.7%, indicating the economic infeasibility of the hybrid system as shown on Figure 13. The profitability of this hybrid PV-wind system is contingent on surplus power being sent to the national grid. The hybrid supply system is expected to generate a yearly cash inflow of $381061.18 as illustrated in Figure 12 by exporting the excess electricity to the grid.

The use of GHG emission to evaluate the technical viability of hybrid RE has seen many phases of studies. Since the global trend of renewable adoption is sustained by the reduction of fossil fuel usage, it is critical for researchers to use GHG emission parameters to evaluate the technical or environmental friendliness of any hybrid RE system. Researchers such as [29]; performed a techno-economic feasibility study of hybrid systems incorporating Diesel generator for a City in United Arab Emirates. The study reported a 69.6% reduction in GHG emission which is equivalent to 682279 gallons of diesel fuel not used. Again, [30] performed similar economic and technical assessment of HES for several communities in Yemen. In this paper, the environmental assessment revealed a GHG emission reduction of 100% for the solar PV and wind energy system whiles a hybrid solar PV, Wind and diesel generator reduced GHG emission by 97.44 %. If Ghana had a GHG reduction credit rate (say \$25/tCO2 emission), then the proposed system will generate a yearly GHG reduction revenue of \$4096.00. This GHG policy would encourage companies to direct efforts and procedures to reduce their GHG emission footprint. This will cumulatively accelerate the National objective of increasing the RE percentage in its energy mix.

5. Conclusion

The increasing focus on sustainability has prompted communities to seek alternatives to traditional fossil fuel-based electricity generating. Renewables help in the modernization of power system operations and can be economical, particularly in nations where renewable resources are abundant. As a result, the current study investigated the economic and technical viability of a solar PV and wind hybrid system to serve the yearly commercial energy requirement of KMA Office in Kumasi, Ashanti region of Ghana. The viability of the designed hybrid system reveals that, when compared to an off-grid solar PV or wind systems, the hybrid system satisfies the energy requirement of the KMA offices better.

The research also highlighted the significant possibility of exporting the system's excess electricity to the available large grids. Furthermore, the designed hybrid system offers an energy cost of 0.671 \$/kWh, which is somewhat higher than the cost of electricity in Ghana, regarding the high interest rate in the country as compared to other countries with lower lending and discount rate.

The following recommendations are aimed at future research into hybrid renewable energy system: the viability of including a battery storage device to store extra generated energy for other parallel applications. It is also recommended that a comparative analysis be undertaken to maximize economic feasibility while reducing the impact of the intermittent nature of renewable energy resources. While considering coupling to the grid, further studies should ponder on the issue of voltage and frequency stability and reliability of grid connected RETs.

References

[1] IEA 2021 *Key World Energy Statistics 2021* (Paris: OECD)
[2] IEA 2015 *World Energy Outlook 2015 Factsheet* (Paris)
[3] Vohra K, Vodonos A, Schwartz J, Marais E A, Sulprizio M P and Mickley L J 2021 Global mortality from outdoor fine particle pollution generated by fossil fuel combustion: Results from GEOS-Chem *Environmental Research* **195**
[4] Qazi A, Hussain F, Rahim N A B D, Hardaker G, Alghazzawi D, Shaban K and Haruna K 2019 Towards Sustainable Energy: A Systematic Review of Renewable Energy Sources, Technologies, and Public Opinions *IEEE Access* **7** 63837–51
[5] REN21 2020 *Renewables 2020 Global Status Report* (Paris)
[6] REN21 2017 Ren 21 - Renewable Global Futures Report. Great debates towards 100 % renewable energy *Ren21* (Paris)
[7] REN21 2016 *Renewables 2016 Global Status Report* (Paris)
[8] Energy Commission of Ghana 2006 Strategic National Energy Plan: Electricity *Energy Commission* (Accra)

[9] Kemausuor F, Obeng G Y, Brew-Hammond A and Duker A 2011 A review of trends, policies and plans for increasing energy access in Ghana *Renewable and Sustainable Energy Reviews* **15** 5143–54

[10] Acakpovi A and Asabere N Y 2017 Exploring the Fundamentals of Solar Photovoltaic Technology and its Modelling *Renewable Energy Systems* ed S A Kale (New York, USA: Nova Science Publishers) pp 1–37

[11] Hammad B, Al-Sardeah A, Al-Abed M, Nijmeh S and Al-Ghandoor A 2017 Performance and economic comparison of fixed and tracking photovoltaic systems in Jordan *Renewable and Sustainable Energy Reviews* **80** 827–39

[12] Smruti Ranjan Pradhan, Saurav Kumar Singh, Mantu Kumar Choudhury and Dibyalok Dwivedy 2015 Study of Cost Analysis and Emission Analysis for Grid Connected PV Systems using RETSCREEN 4 Simulation Software *International Journal of Engineering Research and* **4**

[13] Al-Soud M S 2018 Implementation of photovoltaic energy generation to reduce the cost of electricity in the water sector in Jordan *International Journal of Energy Technology and Policy* **14** 390–408

[14] Kumi K 2012 *Technoical and Economic Analysis of a 1 MW Grid-Connected Solar* (Kwame Nkrumah University of Science and Technology (KNUST))

[15] Al-Masri H M K, Al-Quraan A, AbuElrub A and Ehsani M 2019 Optimal coordination of wind power and pumped hydro energy storage *Energies (Basel)* **12** 1–15

[16] Khraiwish Dalabeeh A S 2017 Techno-economic analysis of wind power generation for selected locations in Jordan *Renewable Energy* **101** 1369–78

[17] Ammari H D, Al-Rwashdeh S S and Al-Najideen M I 2015 Evaluation of wind energy potential and electricity generation at five locations in Jordan *Sustainable Cities and Society* **15** 135–43

[18] Acakpovi A, Adjei P, Nwulu N and Asabere N Y 2020 Optimal Hybrid Renewable Energy System: A Comparative Study of Wind/Hydrogen/Fuel-Cell and Wind/Battery Storage *Journal of Electrical and Computer Engineering* **2020**

[19] Al-Ghussain L, Samu R, Taylan O and Fahrioglu M 2020 Techno-Economic Comparative Analysis of Renewable Energy Systems: Case Study in Zimbabwe *Inventions 2020, Vol. 5, Page 27* **5** 27

[20] Acakpovi A, Adjei P, Asabere N Y, Sowah R A and Sackey D M 2021 Techno-Economic Evaluation of Hydrogen Fuel Cell Electricity Generation Based on Anloga (Ghana) Wind Regime *https://services.igi-global.com/resolvedoi/resolve.aspx?doi=10.4018/IJEOE.2021070103* **10** 47–69

[21] Halawani M and Özveren C S 2015 A hybrid power system for remote areas in Jordan *Proceedings of the Universities Power Engineering Conference* vol 2015-November (IEEE)

[22] Okonkwo E C, Okwose C F and Abbasoglu S 2017 Techno-Economic Analysis of the Potential Utilization of a Hybrid PV-Wind Turbine System for Commercial Buildings in Jordan *International Journal of Renewable Energy Research* **7** 908–14

[23] Vick B D and Neal B A 2012 Analysis of off-grid hybrid wind turbine/solar PV water pumping systems *Solar Energy* **86** 1197–207

[24] Alharthi Y Z, Siddiki M K and Chaudhry G M 2018 Resource Assessment and Techno-Economic Analysis of a Grid-Connected Solar PV-Wind Hybrid System for Different Locations in Saudi Arabia *Sustainability 2018, Vol. 10, Page 3690* **10** 3690

[25] Budes F A B, Ochoa G V, Obregon L G, Arango-Manrique A and Álvarez J R N 2020 Energy, Economic, and Environmental Evaluation of a Proposed Solar-Wind Power On-grid System Using HOMER Pro®: A Case Study in Colombia *Energies 2020, Vol. 13, Page 1662* **13** 1662

[26] Mahesh A and Sandhu K S 2020 A genetic algorithm based improved optimal sizing strategy for solar-wind-battery hybrid system using energy filter algorithm *Frontiers in Energy* **14** 139–51

[27] Rashid M, Abujubbeh M and Fahrioglu M 2017 Improving capacity factor of transmission lines by hybridizing CSP with wind *2017 4th International Conference on Electrical and Electronics Engineering, ICEEE 2017* 1–5
[28] Abujubbeh M, Marazanye V T, Qadir Z, Fahrioglu M and Batunlu C 2019 Techno-Economic Feasibility Analysis of Grid-Tied PV-Wind Hybrid System to Meet a Typical Household Demand: Case Study-Amman, Jordan *Proceedings - 2019 IEEE 1st Global Power, Energy and Communication Conference, GPECOM 2019* 418–23
[29] Salameh T, Ghenai C, Merabet A and Alkasrawi M 2020 Techno-economical optimization of an integrated stand-alone hybrid solar PV tracking and diesel generator power system in Khorfakkan, United Arab Emirates *Energy* **190**
[30] Mubaarak S, Zhang D, Liu J, Chen Y, Wang L, Zaki S A, Yuan R, Wu J, Zhang Y and Li M 2021 Potential techno-economic feasibility of hybrid energy systems for electrifying various consumers in Yemen *Sustainability (Switzerland)* **13**

Synthesis of ZnO nanoparticles through a simple wet chemical precipitation method

Gem Taka* **and T D Das**
Department of Basic and Applied Science, National Institute of Technology, Arunachal Pradesh, 791112, India

*Email: gtaka490@gmail.com

Abstract. The research on Zinc oxide nanoparticles has become very important on these days due to its unique property and wide range of applications in all the branches of science. In this study, we synthesized ZnO nanoparticles through a simple and cost-effective wet chemical precipitation approach. The XRD spectra revealed the hexagonal wurtzite structure in the prepared ZnO nanoparticles. The UV-Vis absorption peak of the as - prepared ZnO sample was identified at 301.2 nm and was observed to be blue shifted in comparison to the bulk counterpart. The transmittance analysis of the prepared ZnO sample showed very high transmittance to the visible range of light. The widening of the optical band gap was observed in the preparation ZnO sample. The band gap of the synthesized ZnO sample was found to be 3.7943 eV. The theoretical estimation of particle size in the prepared ZnO sample was performed using the Brus model and the particle size was estimated to be 4.38 nm. This study revealed the potential application of prepared ZnO nanoparticles as transparent electrode in solar cell.

1. Introduction

Nowadays, there are increasing demands for the development of semiconductor nanostructures because of their wide range of applications such as electronic and optoelectronic devices. The properties of the matter depend on the dimension and orientation of the structure at the nanoscale. Zinc oxide (ZnO) has a distinct property of various other semiconductor materials. It exists mainly in a hexagonal wurtzite structure having coordination number four [1]. ZnO is generally n-type semiconductor with a large excitonic binding energy of 60 meV and a broad direct band gap of 3.37 eV at normal temperature [2]. Due to high excitonic binding energy, ZnO has high radiative recombination efficiency in spontaneous emission and a low threshold voltage in stimulated emission [3]. It has excellent pyroectric and piezoelectric capabilities, making it suitable for use as mechanical actuators and piezoelectric sensors [4].

ZnO nanoparticles show promising properties for various device applications like transparent electrode in solar cell, chemical sensor [5-7], humidity sensor [8], UV photo sensor and high power FETs. The literature study revealed a variety of synthesis methods for producing ZnO nanoparticles. Some of them are precipitated [9], hydrothermal methods [10], chemical vapor deposition [11], Chemical synthesis method [12] and precipitation method [13, 14]. Chemical synthesis method is the most widely used out of these approaches due to its easy procedure and reasonable cost.

In this work, we explore the preparation of ZnO nanomaterials through a cost effective wet chemical precipitation method. The XRD spectra revealed the structural property of as-prepared ZnO sample. The optical characterization of ZnO nanoparticles was studied using UV - Vis spectroscopy and the theoretical calculation of nanoparticle size was done using effective-mass model [15].

2. Materials and preparation method

All the chemicals used for precursors and maintaining reaction conditions were of reagent grade. They were used without further purification. The distilled water was used as a solvent in the entire solution.The preparation was followed as per the reported method [2]. 0.8g of sodium hydroxide (NaOH) was added to the beaker containing 200 ml of distilled water to make 0.1M NaOH solution. The solution was then stirred at magnetic stirrer for 30 minutes at 400 RPM, maintaining the temperature of the solution at 50^0C. At separate flask, 5.950g of Zn $(NO_3)_2$. $6H_2O$ was added to the beaker containing 200 ml of distilled water to make it 0.1M $(ZnNO_3)_2$. $6H_2O$ solution. Then it was stirred at the magnetic stirrer for 2 hours at 400 RPM maintaining the temperature of the solution at 60^0C. Four drops of 3% ammonia solution was added to the solution to make the pH of the solution to 12. Now, already prepared 0.1M NaOH solution was added to it drop wise lowering its temperature gradually to the room temperature. The solution was further stirred for 2 hours at 400 RPM at room temperature. The final product of the solution with whitish ppt was filtered, washed with ethanol, dried and grinded. The filtered solution was stored for UV Vis analysis.The possible chemical reactions to form ZnO nanoparticles from the precursors in the solution are as follows [11].

$$NaOH(aq) \rightarrow Na^+ + OH^-$$
$$Zn(NO_3)_2(aq) \rightarrow Zn^{2+} + 2NO_3^-$$
$$Zn^{2+} + 2OH^- \rightarrow Zn(OH)_2$$
$$Zn(OH)_2(aq) \rightarrow ZnO(s) + H_2O$$

Figure 1. Synthesis of ZnO nanoparticles.

3. Result and Discussions

3.1. *X-Ray Diffractometer (XRD) Spectra*

The structural properties of as a prepared ZnO powder sample were studied with XRD using Cu K alpha radiation (k = 1.5406 angstrom). The XRD pattern is shown in the fig.(2). The prominent peaks are obtained at 31.57°, 34.16°, 36.56°, 47.75°, 56.44°, 60.53°, 62.83°, 66.33°, 68.13°, 69.12° corresponding to the planes (100), (002), (101), (102), (110), (103), (200), (112), (201), (004) respectively which corresponds to the hexagonal wurtzite structure of ZnO (JCPDS file [36-1451]).

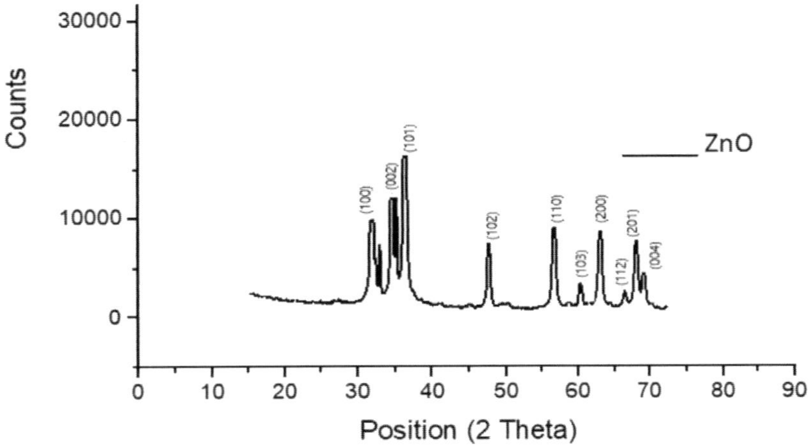

Figure 2. XRD pattern of prepared ZnO nanoparticles.

3.2. UV-Vis absorption study

The optical characterization was performed on UV-Vis spectroscopy having model number JASCO V-730 Spectrophotometer. Wavelengths from 250 nm to 900 nm were evaluated. The room temperature absorption spectrum of as prepared ZnO sample is shown in the fig.(3). The blue shift in absorption peak was observed in prepared ZnO sample in comparison to the bulk counterpart. The sharp absorption peak was identified at 301.2 nm. The quantum size effect, which confirms the formation of nanoparticles, may be responsible for this blue shift in absorbance peak.

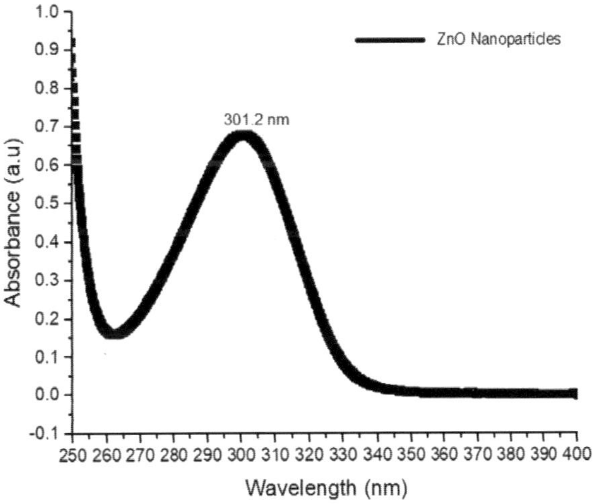

Figure 3. UV-V is absorbance spectrum of ZnO nanoparticles.

The optical transmittance of the prepared ZnO nanoparticles is given in the fig. (4). The transmittance spectrum from 300nm to 900nm was evaluated. High transmittance in the visible region of light is observed.

IOP Conf. Series: Earth and Environmental Science 1042 (2022) 012017 doi:10.1088/1755-1315/1042/1/012017

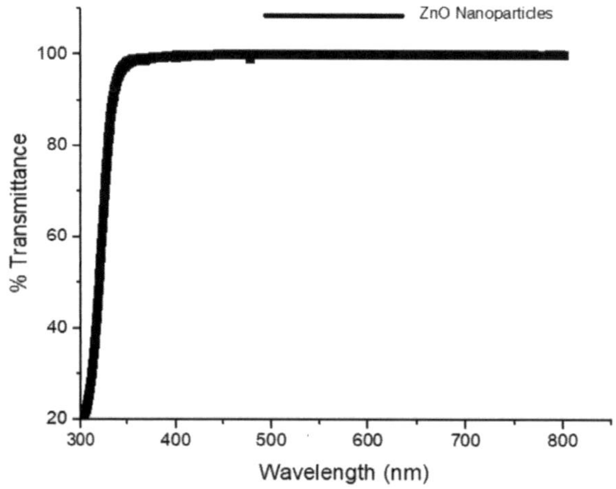

Figure 4. Transmittance curve for ZnO nanoparticles.

The optical band gap can be determined using the Tauc formula [1],

$$(\alpha h\nu)^2 = B\,(h\nu - E_g)$$

Where; α is the absorption coefficient,

$h = 6.626 \times 10^{-34}$ m^2kg/s is Planck's constant,

ν is the frequency of photon, B is the constant and E_g is the band gap energy.

When $(\alpha h\nu)^2 = 0$ then $E_g = h\nu$

Hence, extrapolating the linear portion of the graph $(\alpha h\nu)^2$ Vs $h\nu$ having intercepted on energy axis gives the optical band gap of the sample.

The Tauc plot of as - synthesized ZnO sample is given in the fig.(5). From the Tauc plot, the optical band gap of the as-prepared ZnO sample was found to be 3.7943 eV. Since bulk ZnO has a band gap of 3.37 eV [1], there is band widening in our prepared ZnO sample. This band widening may be attributed to the quantum quenching effect. This confirms the formation of ZnO nanoparticles in our prepared sample.

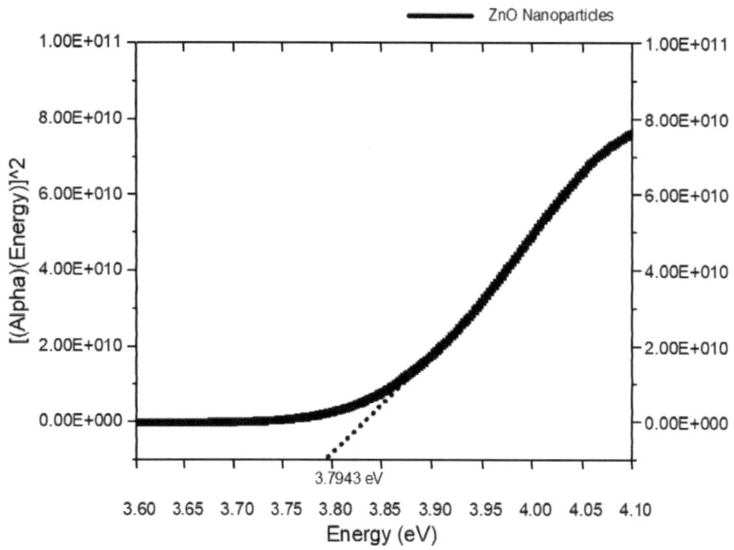

Figure 5. Tauc plot of as prepared ZnO nanoparticles.

3.3. *Theoretical estimation of particle size*

There are large numbers of theoretical model proposed to evaluate qualitative agreement of the predicted dependence of the energy band gap with respect to the particle size obtained from the experimental data of the semiconductor nanoparticles. The most used theoretical model to analyze the dependence of band gap with respect to semiconductor nanoparticle is the "effective mass approximation" or the Brus model [12, 15].

According to the Brus model or effective mass approximation, the energy band gap (E_g) of semiconductor nanoparticle having a spherical shape with a particle radius (r) is given by following equations,

$$E_g(r) = E_{bulk} + \frac{\hbar^2}{8 m_e r^2}\left(\frac{1}{m_e^*} + \frac{1}{m_h^*}\right) - \frac{1.8 e^2}{\varepsilon_o \varepsilon_r r} - 0.248 \frac{4\pi^2 e^4 m_e}{2(4\pi\varepsilon_0\varepsilon_r)^2 h^2 \left(\frac{1}{m_e^*} - \frac{1}{m_h^*}\right)}$$

$$E_g(r) - E_{bulk} = \Delta E_g = \frac{\hbar^2}{8 m_e r^2}\left(\frac{1}{m_e^*} + \frac{1}{m_h^*}\right) - \frac{1.8 e^2}{\varepsilon_o \varepsilon_r r} - 0.248 \frac{4\pi^2 e^4 m_e}{2(4\pi\varepsilon_0\varepsilon_r)^2 h^2 \left(\frac{1}{m_e^*} - \frac{1}{m_h^*}\right)}$$

Where,

E_{bulk} is the bulk energy band gap value of semiconductor material.

h, is a plank constant.

m_e, is electron mass.

m_h^*, is effective mass of hole.

m_e^*, is effective mass of electron;

In our case, $\Delta E_g = 0.4243 eV$

The first term in eq. (2) refers to the quantum localization term that is the kinetic energy term which shifts the bandgap to the higher energies proportional to the negative square of particle radius. The second term in eq. (2) is due to the screened coulombic interaction between the electron and hole which shifts the bandgap to the lower energy proportional to the negative one of particle radius. The third term in eq. (2) is the size independent term which is the salvation energy loss term and it is generally very small and can be neglected.

Putting the corresponding values in equations (2) and solving, the value of r can be evaluated and in our case it is calculated to be 2.19 nm. Hence, the ZnO nanoparticle is estimated to be $2r = 4.32$ nm. Similar result has been observed by other workers like Domenica Tommasa Donia et al [12].

4. Conclusions

In this work, we have synthesized ZnO nanoparticles by employing a simple and cost effective, wet chemical precipitation method. The chemicals, namely, Zinc nitrate hexahydrate and sodium hydroxide have been used to prepare ZnO nanomaterials. The XRD spectra revealed the hexagonal wurtzite structure in the prepared ZnO nanoparticles. The optical characterization has been performed using the UV-Vis spectroscopy. The blue shift in the absorbance spectrum of as - prepared ZnO sample with respect to the bulk counterpart has been identified. The absorption peak of as prepared ZnO sample was centered at 301.2 nm. The widening of band gap has been observed for the prepared ZnO sample. The optical band gap of as prepared ZnO nanoparticle was found to be 3.7943 eV. The transmittance spectra analysis revealed very high transmittance of a prepared ZnO sample to the visible region. The particle size of the synthesized ZnO nanoparticle was estimated using effective mass model and was found to be 4.38 nm.

References

[1] R. Dash, C. Mahender, P. Kumar Sahoo et al. 2021 Preparation of ZnO layer for solar cell application *Materials Today: Proceedings* **41**(2) pp 161-164

[2] J.N. Hasnidawani et al. 2016 Synthesis of ZnO nanostructures using Sol- Gel Method *Procedia Chemistry* **19** pp.211-216.

[3] B Illy et al. 2005 Electrochemical growth of ZnO nanoplates *Nanotechnology* **16**(2) pp 320–324

[4] G. Vijaya Prakash et al. 2008 Fabrication and Optoelectronic Characterisation of ZnO Photonic Structures *Materials Letters* **62** pp 1183–1186.

[5] Savita Chaudhary et al. 2018 Chemical Sensing Applications of ZnO Nanomaterials *Materials* **11** p 287

[6] Ahmed Mishaal Mohammed et al. 2017 Nanostructured ZnO-based biosensor: DNA immobilization and hybridization *Sensing and Bio-Sensing Research* **15** pp 46–52.

[7] S G Onkar et al. 2020 Gas sensing behavior of ZnO thick film sensor towards H_2S, NH_3, LPG and CO_2 *Journal of Physics: Conference Series* **16** p 012060

[8] Guofeng Hu et al. 2014 Piezotronic effect enhanced Schottky-contact ZnO micro/nanowire humidity sensor *Nano Res.* **7** pp 1083-1091

[9] Ajay Kushwaha et al. 2011 Energy Efficient and Size Tunable Synthesis of 'Colloidal' Zinc Oxide Nanoparticles by Simple Beaker Chemistry *AIP Conference Proceedings* **13** pp 391-392

[10] S. F. U. Farhad et al. 2018 Facile synthesis of oriented zinc oxide seed layer for the hydrothermal growth of zinc oxide nanorods *Bangladesh J. Sci. Ind. Res.* **53**(4) pp 233-244.

[11] R. A. Al-Wardy et al. 2021 Optical Properties of ZnO Films Prepared by CBD technique *Journal of Ovonic Research* **17**(1) pp 53 – 60.

[12] Domenica Tommasa 2021 Room Temperature Syntheses of ZnO and Their Structures *Donia et al. Symmetry* **13**(4) p 733.

[13] Talam, S., Karumuri, S.R. and Gunnam, N., 2012. Synthesis, characterization, and spectroscopic properties of ZnO nanoparticles. *International Scholarly Research Notices, 2012.*

[14] M. Schmitt 2015 Synthesis and testing of ZnO nanoparticles for photo-initiation: experimental observation of two different non-migration initiators for bulk polymerization *Nanoscale* **7**(21) pp 9532–9544.

[15] Musa et al. 2019 Study of Optical Energy Gap And Quantum Confinment Effects in Zinc Oxide Nanoparticles and Nanorods *Digest Journal of Nanomaterials and Biostructures* **14** (1) pp.119-125

IOP Conf. Series: Earth and Environmental Science 1042 (2022) 012018 doi:10.1088/1755-1315/1042/1/012018

Innovative design of grid connected solar/diesel hybrid system using RETScreen software

Daniel Kofi[1], Amevi Acakpovi[2*], Patrick Adjei[1], Robert Sowah[3], George K. Aggrey[4], Godfred Tay[1], Mustapha Sulley[1], Maxwell Afonope[1]

[1] Regional Maritime University, Department of Electrical and Electronic Engineering, Accra, Ghana

[2] Accra Technical University, Department of Electrical and Electronic Engineering, Accra, Ghana

[3] University of Ghana, Department of Computer Engineering, Accra, Ghana

[4] University of Cape-Coast, Department of Computer Science and Information Technology, Cape-Coast, Ghana

*Email: acakpovia@gmail.com

Abstract. Accessibility to dependable energy resource is vital to the emerging economy to function appropriately in this contemporary world for both residential and commercial purposes. Technological advancement has opened avenues for more sophisticated technologies to combine multiple energy sources to generate affordable electricity for residential and industrial purposes. The study, therefore aims at analyzing the fiscal benefits of hybrid Solar PV and Diesel Generator (DG) (PV-DG) grid-connected system using RETSceen software. The study focused on a specific location, J. A. Plant Pool Ghana Limited, warehouse Department. The study recommended and proposed an effective design of a hybrid PV-DG grid connected electricity supply for the warehouse Department. The economic viability of the project and the payback period obtained through computation were very attractive. Also, the total revenue of electricity exported to the grid annually and GHG emission reduction were within the standard benchmark. The study finally concurred that renewable energy sources such as solar when hybridized with any other energy systems, lead to a noticeable electrical cost reduction. Therefore, this system design provides multiple benefits, including; improved reliability, reduced emission and significant cost reduction.

1. Introduction

Among extensive varieties of renewable energy, solar PV energy is exclusive and is commonly used in the residential and commercial sectors in Ghana. There are two categories of solar energy in relation to their mode of operation: active and passive solar energy. Active Solar energy are those that operates with other energy conversion measures often involving heating, cooling and advanced mechanical systems, while the passive mechanisms are those without the use of further mechanical energy conversion systems. The two types of active solar energy are the thermal energy and solar Photovoltaic (PV). Solar thermal energy often involves the conversion of heat energy from the sun and involved popularly known technologies such as concentrated solar power (CSP), water heating, etc. whiles the Solar PV converts the radiation from the sun to a useful electricity via photovoltaic cells

[1–3]. A hybrid solar system is an amalgamation of solar PV and other alternate energy resources such as fuel cells, diesel generators, grid, hydropower, wind and many others so as to supply a reliable, un-interrupted power to a load due the intermittency in the solar radiation and also to minimize cost. Most hybrid systems are standalone system operating as off-grid. Grid is a significant scale of power supply or macro-grid; both generate hundreds of Megawatts to hundreds of gigawatts (GW) with a centralized system for distribution and transmission across a country and even beyond.

Hybrid systems are either standalone system or grid connected systems that store and provide power through batteries or diesel generator (DG). Solar-diesel hybrid systems are seen as more reliable and a bit affordable compared to other renewables, especially in Ghana owing to the intense availability of the resource, the existing policy on cost reduction of system components[4–8]. Currently, hybrid systems are more grid connected in Ghana for economic reasons.

Hybrid PV-Diesel generators are becoming a widespread solution for continuous electricity[9,10]. It's therefore the aim of this research paper to analyze the economic gains of a grid-connected hybrid PVDG system using RETScreen software with a specific load: J. A. Plant Pool Ghana Limited, warehouse department so as to recommend the appropriate size of the PV-DG electricity plant for the Warehouse Department. Also, the relevance of this paper is due to the fact that the RETScreen software has really not been used for similar studies previously in Ghana though it seems to promise effective results from the literature. In a similar study, the Federal Government of Nigeria used the RETSceen Expert software from the Canadian government to attract international investors through economic analysis, technical analysis and sensitivity analysis [11].

RETScreen Expert software is a carbon free energy controlling, planning and simulation software for assessing the techno-socio-economic viability and ecological sustainability for the recommended Solar PV DG system for J. A. Plant Pool Ghana Limited. A solar PV and DG technologies were considered during the design stage in this study illustrated in the remaining paragraphs.

Given the capital cost per Watt peak (Wp) for PV compared to DG, the design consisted in ensuring an optimal balance between the PV and the DG, where the PV produced all the electricity that is consumed during the day and in the evening via the grid. The DG only produces electricity when both the grid and the PV are not available. Therefore, the DG served only critical load by day and all the evening load in the event when the PV and the grid are off.

The rest of the paper is structured as follows: section 2 deals with the materials and methods, section 4 presents the results and discusses them, section five presents the conclusion.

2. Materials and Methods

Solar PV is a recognized technology of electricity generation that minimizes CO2 emissions that subsequently lead to global warming and at the same time economic challenges [3,12–14]. The RETScreen software is used to execute environmental and socio-economic assessment of substituting grid with renewable resource for lesser infrastructure and have proven potent. Also, [15] presented the benefit and strengths of using the RetScreen Expert package to design a solar PV system for a specific load.

Pakistan's energy report using the RETScreen software indicates that implementing a 5-kW standalone solar PV system with 0.6–0.7 t_{CO2} falls in GHG emissions in diverse regions of Pakistan under usual climatic condition [10]. Their study established a new potential electricity price-tariff using the RETScreen.

RETScreen can also be used to determine a solar PV system's financial and environmental impact as well as handle other forms of renewables such us wind energy.

In Ghana, RETScreen software has not been used intensively among the available literature despite the promise of reliability and effectiveness. However, a few studies used it to assess the economic viability and potency of PV in Ghana [16]. The 5 MW of grid-connected PV of SPR-320E-WHTD PV module shows the government's contribution to providing incentives relieves and other subsidies to create the investment boost in the energy sector of Ghana to can help curb the recent load shedding and the power outages, thus increasing economic resilience of the energy sector.

IOP Conf. Series: Earth and Environmental Science 1042 (2022) 012018 doi:10.1088/1755-1315/1042/1/012018

2.1. RETScreen Overview

RETScreen is a unique, innovative renewable energy software for planning, education and capacity created to help and support decision making in the development and management of renewable energy by the Canadian Government [17]. The software is an intellectual property of 307 networks of experts and more, including academia, government, industry with partnerships from groups like the Renewable Energy and Energy Efficiency Partnership (REEEP), National Aeronautics and Space Administration (NASA), Global Environment Facility (GEF), United Nations Environment Program (UNEP), Leonardo Energy Initiative, Energy and Environment Foundation, World Bank's Prototype Carbon Fund (PCF). Aside the latest software version; RETScreen Expert is known only to be in a viewer mode for free [8]. The software is written in 36 dialects covering further than two-thirds of the world's populace in over 222 nations and terrains world wide. The uniqueness of the software meaningly reduce the time and cost related parameters when classifying and evaluating possible energy issues and developments that may rise at the planning, feasibility, implementation and engineering stages. The RETScreen PV project model is used universally to compute life-cycle costs, energy production and GHG reduction are the three elementary PV applications: water pumping, on-grid and off-grid. The Expert vision was used for this research.

2.2. *Computational Approach*

The study considered a specific location: the J. A. Plant Pool Ghana Limited (JAPP), on the Alajo link Rd, Accra, Ghana (5°36.3'N, 0°12.4'W) as illustrated in Figure one. The paper was mainly concerned with the design of a 90.2kWh electricity grid-connected hybrid solar PV and Diesel Generator (DG) system to power only the Warehouse Department of JAPP using the RETScreen software.

Figure 1. Google map view of site location

JAPP is a Jospong subsidiary, a leader in supply of luxurious buses, lubricants, heavy-duty trucks, earth moving equipment and tyres in Ghana and outside Ghana for over 12 years.

The vision of the company is to be the industry leader in sales, rental and repair of construction, mining and transport related machinery in Ghana and beyond by providing high quality construction and mining equipment, haulage and transport related services with excellent customer service to increase shareholder value.

The warehouse department is made of enclosed storage area with three offices on the J. A. Plant Pool premises at GA-090-0785, Dzorwulu, Accra. The Warehouse is the busiest place of the company since it houses all the company's stock valued as 2,475kUSD with 81% of availability, 144,743 pieces of Spare parts on 950m2 area. The offices are equipped with laptops, printers, refrigerators, air conditions, telephones and lights.

IOP Conf. Series: Earth and Environmental Science 1042 (2022) 012018 doi:10.1088/1755-1315/1042/1/012018

2.3. Energy Audit

Energy audit was conducted to run a summary on the load profile at the Department to determine the maximum, minimum and critical load for the grid connected solar PV and DG hybrid system using RETScreen software for the simulation and finally for the design. Although the main research study was conducted with the RETScreen software for the purposes of accuracy and efficiency, the design was transposed on the Homer software to conduct performance comparison and draw meaningful conclusions. For instance, the average annual solar radiation of 4.94kW/m2/day was used in the calculation of the inverter; the array area and capacity were obtained from Homer software. Homer software is another efficient energy software mostly used for energy optimalization and all other energy socio-economic analysis. The load profile of JAPP is presented in Table 1.

Table 1. Energy profile at JAPP Warehouse Department in Accra after the energy audit.

LOAD	Rate (*kW*)	Quantity	Usage	Hour (*h*)	Energy (*kWh*)
Laptop	0.02	5	08:00 - 17:00	9	0.9
Colour TV	0.045	1	08:00 - 17:00	9	0.405
Refrigerator	0.035	3	18:00 - 06:00	24	25.0
Light (Indoor)	0.015	6	08:00-17:00	9	0.81
Light (Warehouse)	0.04	12	08:00-17:00	9	4.32
Light (Outdoor)	0.04	4	18:00-06:00	13	2.08
AC	1.44	3	08:00 - 17:00	9	38.88
Telephone	0.002	4	08:00 - 17:00	9	0.072
Printer	0.03	1	08:00 - 17:00	9	0.27
UPS	0.42	1	00.00 - 24.00	24	10.08
TOTAL					**83.02**

2.4. Critical Loads

The design considered some loads to be critical based on their need to operate without interruption for effective running of business. These include laptops, indoor and outdoor lights, UPS, etc. as summarised in Table 2. They are therefore considered as critical loads and are connected directly to the Diesel Generator (DG) having a bypass to the Solar PV and the Grid; hence, in the absence of these two main sources that is the solar PV and the Grid, the DG will automatically supply the needed power to only the critical load to make the design more economical because the RETScreen simulation shows that DG is the more expensive comparing it to the Solar PV and the Grid. With this the electricity supply from the DG is at 10.5kWh.

Table 2. Critical Load

LOAD	Rate (*kW*)	Quantity	Usage	Hour (*h*)	Energy (*kWh*)
Laptop	0.02	5	08:00 - 17:00	9	0.9
Light (Indoor)	0.015	6	08:00-17:00	9	0.81
Light (Outdoor)	0.04	4	18:00-06:00	13	2.08
Telephone	0.002	4	08:00 - 17:00	9	0.072
Printer	0.03	1	08:00 - 17:00	9	0.27
UPS	0.42	1	00.00 - 24.00	24	10.08
TOTAL					**10.15**

2.5. *Low Loads*

These loads are considered as minimal load and for the purpose of this design they are mostly required in the evening. Generally, their power ratings are far lower than the critical load. Consequently, it is foreseen that the proposed PV-DG hybrid system will not fail because the capacity of the DG will be able to supply the needed load when the Solar PV and the Grid are off. A summary of the low loads is presented in Table 3.

Table 3. Low Load

LOAD	Rate *(kW)*	Quantity	Usage	Hour *(h)*	Energy *(kWh)*
Light (Outdoor)	0.04	4	18:00-06:00	13	2.08
UPS	0.42	1	00.00 - 24.00	24	10.08
TOTAL					**7.12**

2.6. *Design Profile with RETScreen software*

The location setting for the load profile using the RETScreen software has been stated in Figure two. The parameters and quantities have also been stated at Table 4 and 5 for the photovoltaic and the reciprocating engine (Diesel Generator) respectively. With the appropriate and successfully input of the parameters into the RETscreen Expert Software for the design proposed, run the simulation using the tools on the menu bar and generate the report for the design.

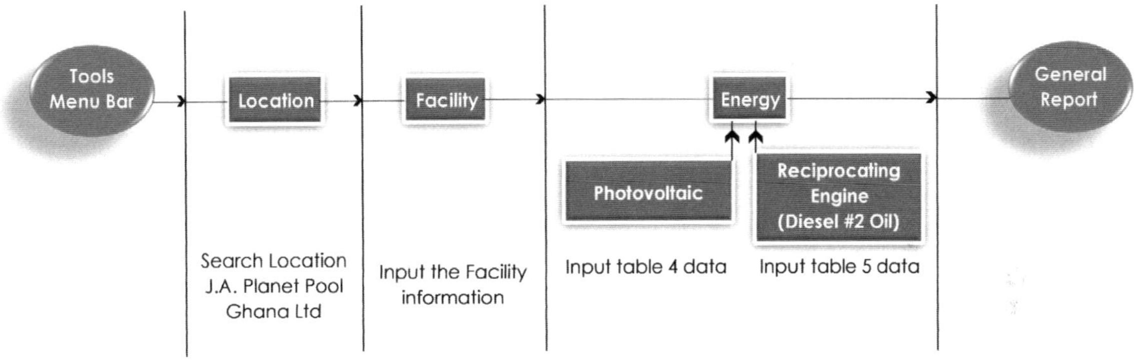

Figure 2. Breakdown of design procedure

Table 4. Solar Photovoltaic input parameters

Parameters	Values
Power capacity	0.01 kW
No. of units	500.00
Capacity factor	91.00%
Initial cost	2100.00$/kW-year
O&M cost (saving)	25.00$/kW
Electricity export rate	0.10$/kWh
Electricity exported to grid	0.08MWh
Electricity export revenue	7.97$

Table 5. Diesel Generator (DG) input parameters

Parameters	Values
Fuel type	Diesel (#2 oil) - L
Fuel rate	0.80$/L
Power capacity	0.004 kW
Availability	30%
Heat rate	13000.00kJ/kWh
Fuel requires	0.0001GJ/h
Initial cost	900.00$/kW
O&M cost (saving)	149.00$/kW-year
Fuel consumption	3.6L
Fuel cost	2.85$
Electricity export rate	0.10$/kWh
Electricity exported to grid	0.01MWh
Electricity export revenue	1.05$

3. Results and Discussion

The energy source from the PV panels is considered as the main electricity supply. The PV supply the power during the day and in the evenings or bad weather periods the electricity is supply by the grid, the diesel generator only supplies electricity in the absents of both the Solar PV and Grid.

3.1. Generation Summary

The RETScreen Expert Software is silent on inverter and its size to convert the DC supply from the Solar PV unlike the Homer software. It therefore important to calculate the inverter parameters, PV array area and the total capacity using the parameters from Homer software.

$$PAC = \frac{Daily\ Load\ \left(kW/d\right)}{RE\ Window\ \left(h/d\right)} \quad (1)$$

$$PAC = \frac{83.02}{10} = 8.302$$

The AC power is converted to DC by the interior with 90% efficiency. This implies the DC power obtained will be as shown below.

$$PDC = \frac{PAC}{Inv_{eff}} \quad (2)$$

$$PDC = \frac{8.302}{0.9} = 9.22\ kW$$

The DC power is not used due to system and other losses. Therefore, DC was adjusted to 40% as shown below.

$$PDC_{adjusted} = 40\% PDC + PDC \tag{3}$$

$$PDC_{adjusted} = 0.4 \times 9.22 + 9.22 = 12.91\ kW$$

Subsequently, the PV array area is determined as

$$A = \frac{PDC_{adjusted}}{1\left(kW/m^2\right) \times Eff_{PV}} \tag{4}$$

With the PV system's efficiency being 0.15, A is estimated as follows

$$A = \frac{12.91}{1 \times 0.15} = 84.067 m^2$$

Estimating the PV area with the system total capacity give us

$$G = E_t \times A \tag{5}$$

Where G is the radiation from the sun and ET is the total energy. With $E_T = 4.94\ and\ G = 84.067m^2$:

$$G = 4.94 \times 84.067 = 415.28\ kWh$$

Therefore, the system's capacity is determined as follows

Table 6. shows, the energy summary show that the initial cost for this project is \$24.60. The 9 units of electricity generated is sold to the grid and payback period is 4.6 years. The operation and the maintenance cost are only \$0.85 which make it more economical.

$$C = \frac{G}{RE_{window}} \tag{6}$$

$$C = \frac{415.28}{10} = 41.528 \approx 42 \text{ kW}$$

Table 6. Energy summary

Electricity exported to grid	Capacity *(kW)*	Electricity *(kWh)*	Initial Cost *($)*	Electricity export revenue *($)*	Fuel Cost *($)*	O&M Cost (savings) *($)*	Single paybac *(yr)*
Photovoltaic	0.001	79.7	21	8	0	0.25	2.7
Reciprocating Engine (DG)	0.004	10.5	3.6	1.3	2.9	0.6	None
Total	**0.01**	**90.2**	**24.6**	**9.3**	**10.08**	**0.85**	**4.6**

3.2. *Financial Analysis*

The economic life span of the project is 20 years with a payback period is 4.6 years as shown in Figure three. The unit cost of electricity is \$0.076kWh.

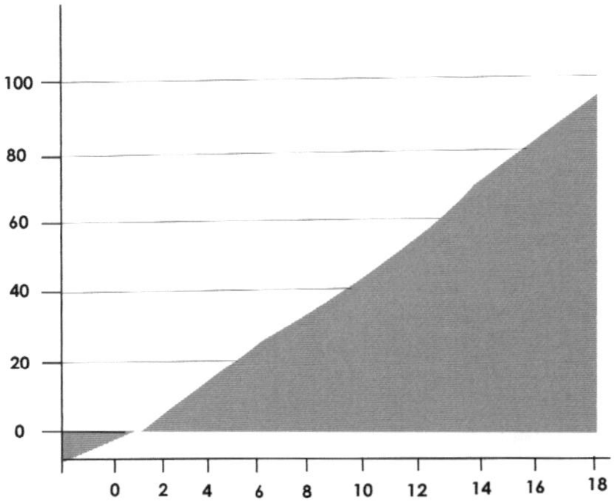

Figure 3. Payback period

3.3. GHG Emission

The gross annual GHG emission reduction is 78% against the standard as illustrated in Figure four. This supports the assertion that the proposed design is environmentally friendly and therefore more accepted.

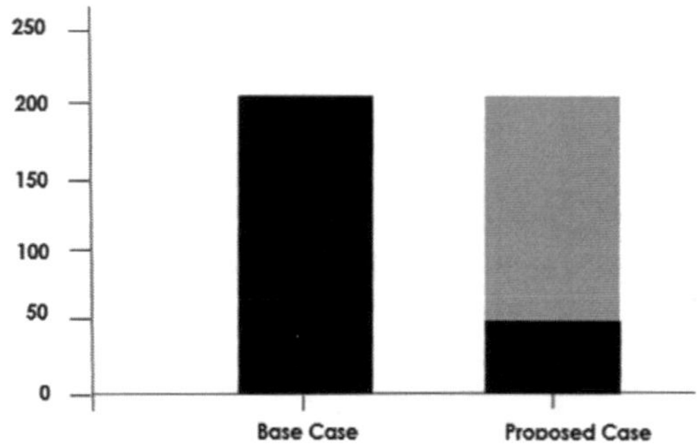

Figure 4. GHG Emission

4. Conclusion

The findings of the study corroborate previous studies regarding the benefits of the hybrid grid connected power systems with solar resources. A few of previous study supporting the assertion of immense benefits in terms of emission and cost reduction while hybridizing solar with other renewables, especially in West Africa are: [18–24]

A detail list of benefits suggested by most of these studies includes: cost reduction, a guarantee of continuous power supply, reduction of pollution and emissions.

The simulation of the load profile with RETScreen Expert software for the 90.5kWh grid-connected solar photovoltaic-diesel hybrid system with the constituent generating capacity of 79.7kWh from the solar PV and 10.5kWh from the Diesel Generator respectively shows that the supply efficiency and the reliability greatly improved. Also, an economic viability established with the fact that the project lifespan is 20 years with an approximate Four year and six months payback period.

The system is designed and sized to export the unused generated electricity to the grid. A total of 0.9 MWh is exported to the grid annually at a revenue of $9.02 with GHG emission reduction of 0.03 representing 78% per annum as against the global benchmark and standard.

The levelized cost of electricity per unit is $0.076 kWh, which is equivalent to Gh₵6.00 per unit cost of electricity in Ghana cedis. Comparatively is far cheaper than per unit cost of Gh₵24.2 for non-residential load [3] in Ghana.

Therefore, this study concurs that hybrid energy systems with renewable energy sources like solar, contribute to an obvious cost savings on electricity. It can therefore be concluded that the Grid Connected Solar PV/Diesel Hybrid system provides multiple benefits including; improved reliability, continuous power supply, reduced emissions and pollution, good operational life and reduced cost.

References

[1] Acakpovi A 2016 Basics Of Solar Photovoltaic Technology

[2] Acakpovi A Technical Training as A Key for a Fast and Effective Industrialization in Ghana 1–4

[3] Acakpovi A and Asabere N Y 2017 Exploring the Fundamentals of Solar Photovoltaic Technology and its Modelling *Renewable Energy Systems* ed S A Kale (New York, USA: Nova Science Publishers) pp 1–37

[4] Acakpovi A 2016 Original Framework for Optimizing Hybrid Energy Supply *Hindawi Publishing Corporation, Journal of Energy* **2016** 1–11

[5] Negi S and Mathew L 2014 Hybrid Renewable Energy System : A Review *International Journal of Electronic and Electrical Engineering* **7** 535–42

[6] Szumanowski A 2013 *Hybrid electric power train engineering and technology: Modeling, control, and simulation*

[7] Acakpovi A, Hagan E ben and Fifatin F X 2015 Cost Optimization of an Electrical Energy Supply from a Hybrid Solar, Wind and Hydropower Plant *International Journal of Computer Applications (IJCA)* **114** 44–51

[8] Acakpovi A, Hagan E ben and Fifatin F X 2015 Cost Optimization of an Electrical Energy Supply from a Hybrid Solar, Wind and Hydropower Plant *International Journal of Computer Applications* **114** 975–8887

[9] Acakpovi A, Michael M B and Majeed I B 2017 Grid connected hybrid solar and diesel generator set: A cost optimization with HOMER *Advances in Science, Technology and Engineering Systems* **2** 92–8

[10] Acakpovi A, Adjei P, Nwulu N and Asabere N Y 2020 Optimal Hybrid Renewable Energy System: A Comparative Study of Wind/Hydrogen/Fuel-Cell and Wind/Battery Storage ed P-F Pai *Journal of Electrical and Computer Engineering* **2020** 1–15

[11] Energy T O C and Analysis P 2005 *RETScreen ® International Clean Energy Project Analysis: RETScreen ® Engineering & Cases Textbook* (Canada: Ministry of Natural Resources Canada)

[12] Şen Z 2008 *Solar energy fundamentals and modeling techniques: Atmosphere, environment, climate change and renewable energy*

[13] Verma D, Nema S, Shandilya A M and Dash S K 2016 Maximum power point tracking (MPPT) techniques: Recapitulation in solar photovoltaic systems *Renewable and Sustainable Energy Reviews* **54** 1018–34

[14] Acakpovi A, Michael M B, Asabere N Y and Honvo J 2017 Exploring the Fundamentals of Solar Photovoltaic Technology and its Modelling *Renewable Energy Systems* ed S A Kale (New-York: NOVA) pp 1–282

[15] Leng G J 1999 RETScreen TM International: A Standarized Tool for Assessing Potential Renewable Energy Projects 1–4

IOP Conf. Series: Earth and Environmental Science 1042 (2022) 012018 doi:10.1088/1755-1315/1042/1/012018

[16] Acakpovi A, Xavier F F and Awuah-baffour R Design Of Analytical Method Of Sizing Photovoltaic Water PVP system installations require many investigations in theories in order to determine of the best size of each components . Indeed , a bad sizing could create unnecessary expenses . Reducing the co 1–18

[17] Ganoe R E, Stackhouse P W and Deyoung R J 2014 *RETScreen® Plus Software Tutorial* (Hampton, Virginia)

[18] Okere A and Tariq Iqbal M 2021 Techno-economic Comparison of Emerging Solar PV modules for Utility Scale PV installation *2021 IEEE 12th Annual Information Technology, Electronics and Mobile Communication Conference, IEMCON 2021* 827–33

[19] Atluri K, Hananya S M and Navothna B 2018 Performance of Rooftop Solar PV System with Crystalline Solar Cells *2018 National Power Engineering Conference, NPEC 2018*

[20] Ayodele E, Misra S, Damasevicius R and Maskeliunas R 2019 Hybrid microgrid for microfinance institutions in rural areas – A field demonstration in West Africa *Sustainable Energy Technologies and Assessments* **35** 89–97

[21] Conway D, Robinson B, Mudimu P, Chitekwe T, Koranteng K and Swilling M 2019 Exploring hybrid models for universal access to basic solar energy services in informal settlements: Case studies from South Africa and Zimbabwe *Energy Research & Social Science* **56** 101202

[22] Maatallah T, Ghodhbane N and ben Nasrallah S 2016 Assessment viability for hybrid energy system (PV/wind/diesel) with storage in the northernmost city in Africa, Bizerte, Tunisia *Renewable and Sustainable Energy Reviews* **59** 1639–52

[23] Hohne P A, Kusakana K and Numbi B P 2019 Optimal energy management and economic analysis of a grid-connected hybrid solar water heating system: A case of Bloemfontein, South Africa *Sustainable Energy Technologies and Assessments* **31** 273–91

[24] Aderemi B A, Daniel Chowdhury S P, Olwal T O and Abu-Mahfouz A M 2018 Techno-Economic Feasibility of Hybrid Solar Photovoltaic and Battery Energy Storage Power System for a Mobile Cellular Base Station in Soshanguve, South Africa *Energies 2018, Vol. 11, Page 1572* **11** 1572

IOP Conf. Series: Earth and Environmental Science 1042 (2022) 012019 doi:10.1088/1755-1315/1042/1/012019

Economic-environmental impacts on reduction of energy consumption in technical high schools during the COVID-19 pandemic: case of Manzanillo, Mexico

V D Cuevas-Rodríguez[1], G M Rodríguez-Torres[2], M Morales-Máximo[1], L B López-Sosa[1,*]

[1]Programas Académicos de Posgrado, Universidad Intercultural Indígena de Michoacán, Carretera Pátzcuaro-Huecorio Km3, Pátzcuaro C.P. 61614, México
[2]Universidad Tecnológica de la Construcción, Nicolás Ballesteros 1200. Ciudad Industrial 4a Etapa 58200. Morelia, Michoacán

*Email: lbernardo.lopez@uiim.edu.mx

Abstract. A study is presented on the economic and environmental impacts, due to the reduction of electricity consumption in a Technical High School in Manzanillo, Mexico, during the COVID-19 pandemic, as well as the impacts generated by the reduction of student mobility. This research shows an analysis of the reduction of electrical energy consumption, the reduction of monthly economic costs and CO_2 emissions in the studied institution, these impacts on student mobility during confinement between November 2019 and October 2021 are also analyzed. Results show electrical energy savings of approximately 60MWh during this period, about 34 $TnCO_2$ and an economic saving of 8654 USD. It has also estimated an economic saving in student mobility of approximately 312,840 USD and a reduction in emissions of the order of 64 $TnCO_2$. This research also suggests the implementation of technologies that use renewable energy sources as a strategy in the face of this health emergency and contributes to raising future scenarios for sustainable energy planning.

1. Introduction

There are multiple impacts generated by the COVID-19 pandemic [1]. Until now, it has not been possible to quantify all the effects on the different sectors of the population, from the health area to the educational sector [2], the latter being one of the ones that suffered the most changes in its teaching-learning practices. In particular, in the presence of Covid-19 in Mexico, the education sector will have to face new challenges: first, return to face-to-face activities with an alternative hybrid system, and second, face the training processes for students and teachers to face new health emergencies that allow the incorporation of information and communication technologies [3] and move towards new post-pandemic processes. In the energy sector, educational institutions have faced various changes during the pandemic [2,4,5], which are manifested in cost savings due to energy consumption and reduction of CO_2 emissions. The confinement in the educational sector generated great changes in the daily life of students, teachers, and educational services.

IOP Conf. Series: Earth and Environmental Science 1042 (2022) 012019 doi:10.1088/1755-1315/1042/1/012019

This research addresses the case study of the estimation of the economic and environmental impacts, a product of the reduction of the consumption of electrical energy and mobility in a Technical High School in the city of Manzanillo, in western Mexico. The results of this research can be used to develop energy efficiency and carbon footprint reduction programs in the education sector, in addition, at the national level this study is extrapolated to show the impacts in all technical high schools and encourage reflection on the transition to new energy production and consumption schemes in the education sector, after the COVID-19 pandemic.

2. Materials and Methods

This research was carried out in the city of Manzanillo, in the state of Colima (19 ° 03′08 ″ N 104 ° 18′57 ″ W), in western Mexico (Figure 1), in a Technical High School. This institution had a student enrollment of 742 in 2019-20202, 711 in 2020-2021, and 633 at the end of 2021. The electrical energy consumption data used in this study is from the period of November 2019 to October 2021, Before and during the pandemic (2019-2020-2021), this information was obtained from the Federal Electricity Commission (CFE is Spanish), which is the institution in Mexico that is responsible for providing electrical energy in the residential, educational and of services. Data on the cost of electricity was also obtained from the CFE. To estimate CO_2 emissions, the emission factor of the Energy Regulatory Commission of 0.55 $KgCO_2e$ / kWh [6] was used. A strategy is also proposed for the use of the economic resources that have been saved during the pandemic by reducing the consumption of electrical energy, implementing solar lamps of 48 USD, and lighting of 10 square meters.

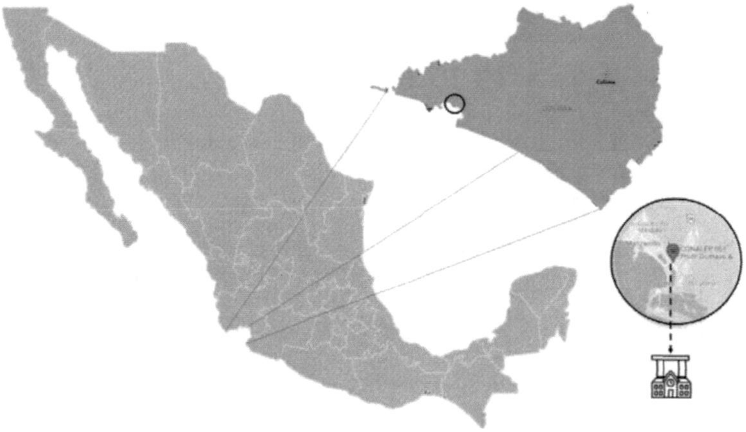

Figure 1. Spatial analysis of the case study

Student mobility is carried out in public passenger-cargo vehicles with a capacity of 30-40 students. To estimate fuel consumption due to mobility, a mobility scenario of a radius of 10 km was considered, which is the distance that most students travel on public transport. The cost of the transfer is 1UDS / day, and the performance of the vehicles was considered 4.5 km / l of diesel fuel. The emission factor of 2.59$KgCO_2$ / liter of diesel was used, according to the national inventory of emissions of the Institute of Ecology and Climate Change of Mexico [7]. The results of electricity saving, emission reduction, and cost reduction during confinement due to the COVID-19 pandemic were estimated in an integrated way; These results were extrapolated to the total number of Technical High Schools that the country has (313), to make a preliminary estimate of the impacts at the national level in this type of institutions. Finally, the total economic savings and the total reduction of emissions from mobility and electricity consumption were studied, carrying out a comparative analysis of these results.

3. Results and discussion

In Mexico, compulsory confinement and school closings were carried out as of March 2020. In April, it is common for activities to be suspended in all schools in the country due to the "Holy Week" holidays, therefore The effects of the pandemic on the reduction of electrical energy were observed until May 2020. Since May 2020, administrators, teachers, and students suspended all their activities, and decreasing curves were reported between October 2020, and April 2021 (Figure 2a), these variations are due to the extreme confinement measures in the face of the wave of infections in 2020, and in 2021 due to the Easter holidays again. In May 2021, it was suggested by the educational authorities of the Mexican government to return to face-to-face activities, and this implied an increase in electricity consumption with the gradual return to classes, but in in June, the summer holidays and the end of the school year again reduced consumption due to the absence of people in high schools and at all educational levels.

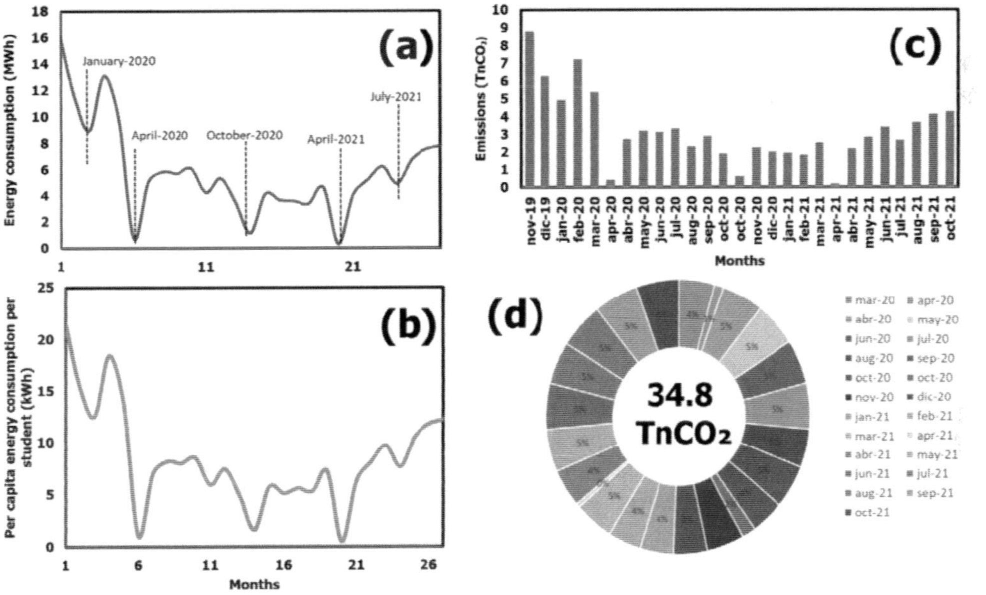

Figure 2. (a) variation in electricity consumption (b) variation in energy consumption per capita per student (c) emissions during the 2019-2021 pandemic (d) total emission estimate.

The behavior of electricity consumption during the pandemic in this high school can also be estimated in terms of per capita demand, which depends on the total number of students in each institution. In this case, the demand can be seen in Figure 2 (b). The variation in energy consumption translates into variation in CO2e emissions, Figure 2 (c) shows the variation in emissions, whose reduction in total compared to the average months of 2019 represents 34.8 TnCO2e (Figure 2d). On the other hand, the percentage of monthly electric power reduction compared to previous years before the pandemic can be seen in Figure 3 (a). Figure 3 (b) shows the monthly economic savings derived from the reduction of electrical energy.

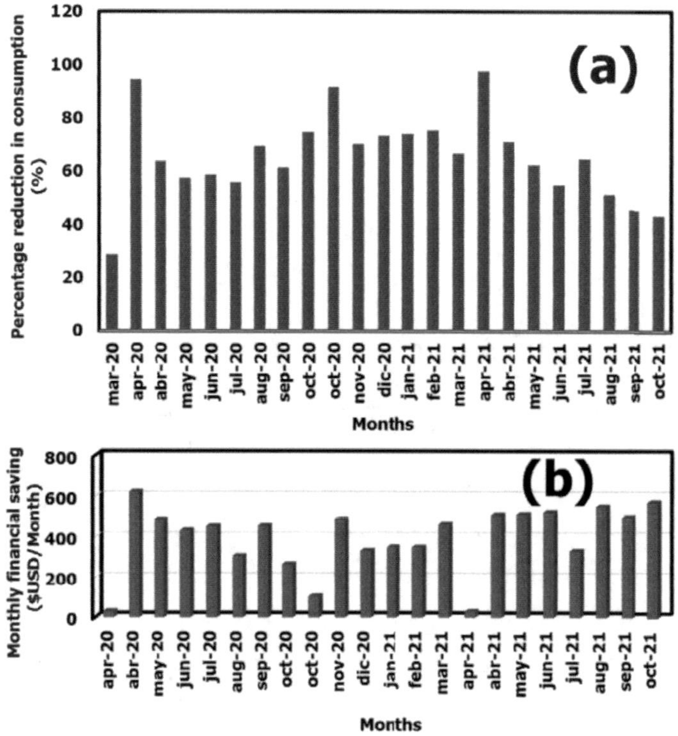

Figure 3. (a) reduction of monthly electricity consumption (b) economic savings during the 2019-2021 pandemic.

The consumption of electrical energy depends on the available infrastructure and the number of students that each educational institution has, since the services depend on the per capita demand and the number of students. The electricity consumption per capita per student between 2019 and 2021 is shown in figure 4 (a), this reduction in electricity consumption during the pandemic has been shown as monthly economic savings in figure 3 (b); This economic saving could be used to implement solar lamps as we have shown in recent research [4], monthly the number of lamps that could be implemented is shown in figure 4 (b), which during the months of study would add up to a total of 56433, equivalent to 56 hectares of lighting.

Confinement and distance classes drastically reduced student mobility. Figure 4 (c) shows the cumulative economic savings for all student enrollment of the technical high school studied. Also, Figure 4 (d) shows the cumulative emission reduction until October 2021. As can be inferred, the reduction in mobility and electrical energy consumption was significant during the COVID-19 pandemic, however, it can be noted that mobility is one of the sectors that generated the greatest economic and emission reductions during this health emergency. Figures 5 (a) and 5 (b) show in a comparative way the percentage of economic reduction and emissions during the pandemic concerning mobility and electricity consumption. In an integrated way for this case study, the environmental, economic and energy impact can be observed in Figure 5 (c); This study can be extrapolated to the total number of high schools in the country, 313, and its impacts can be seen in Figure 5 (d).

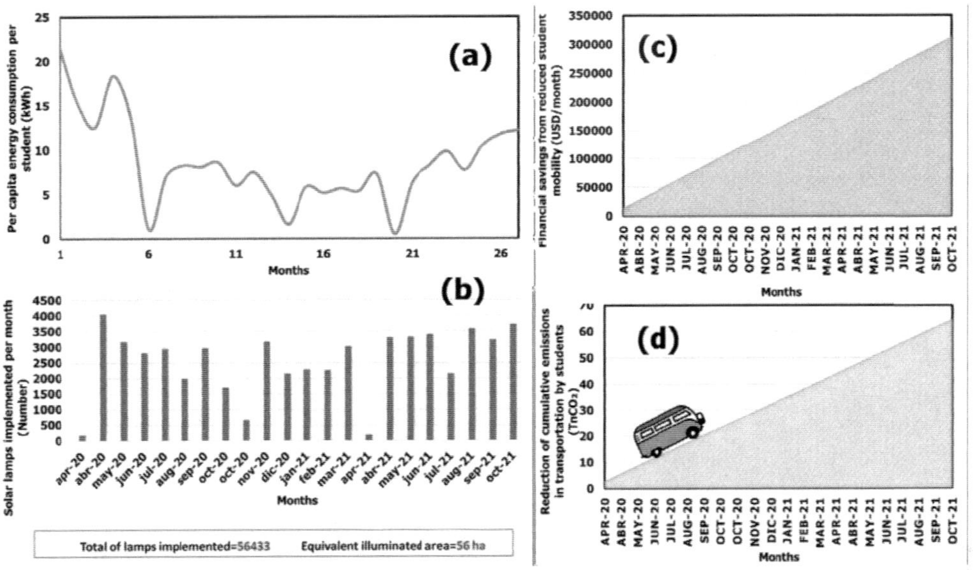

Figure 4. (a) Energy consumption per capita per student (b) proposal for the implementation of solar lamps monthly (c) Economic savings accumulated by the reduction of student mobility (d) reduction of emissions accumulated by the reduction of student mobility.

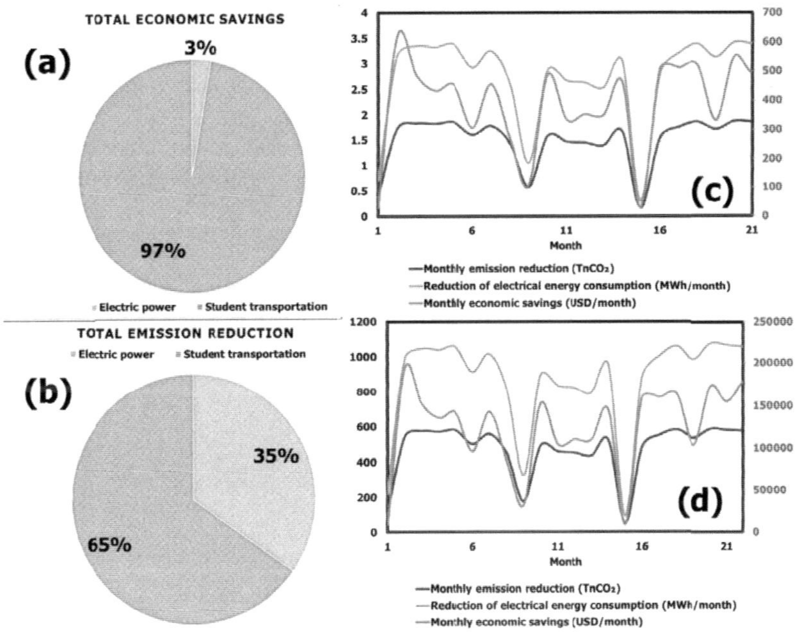

Figure 5. Comparative total savings: (a) economic (b) emissions. Comprehensive analysis: (c) of the technical high school in Manzanillo (d) extrapolated to all schools in Mexico.

In summary, this study has shown the environmental, economic, and energy impacts caused by confinement during the COVID-19 pandemic, in terms of electricity consumption and the reduction of student mobility at the level of a campus in the City of Manzanillo, Mexico, and extrapolated to all schools of the type of high school studied as a case study. The scope of this research allows us to analyze consumption patterns before and during the pandemic, and propose energy transition scenarios from the education sector. As well as looking for alternatives for the insertion of technologies, considering that lighting is one of the main activities of electricity consumption, also that mobility must change and seek to be more and more efficient.

4. Conclusions

This research has shown the economic savings, the magnitude of the reduction of CO2 emissions, as well as the reduction of energy consumption, associated with the electrical energy consumed in a Technical High School, and the reduction of student mobility. Transport is the sector with the highest demand for energy, the highest cost, and the generation of emissions so that during the COVID-19 pandemic the greatest savings were generated with the reduction of mobility. On the other hand, it is important to mention that the extrapolation of this case study, to all the technical high schools in the country, can estimate economic savings of the order of thousands of dollars, reduction of hundreds of tons of CO2 and hundreds of MWh. This research will serve to establish carbon footprint reduction plans in Mexican high schools, as well as strategic energy efficiency programs after the pandemic in this educational sector.

References

[1] Zheng J. SARS-CoV-2 : an Emerging Coronavirus that Causes a Global Threat 2020;**16**.
[2] Shahzad A, Hassan R, Aremu AY, Hussain A, Lodhi RN. Effects of COVID - 19 in E - learning on higher education institution students : the group comparison between male and female. Qual Quant 2021;**55**:805–26.
[3] A. F-PMR& N-CC. Diagnóstico de necesidades de capacitación en el uso de plataformas virtuales ante la contingencia del COVID-19 en los estudiantes y docentes de Educación Media Superior Tecnológica. Rev Dilemas Contemp Educ Política y Valores 2020;17:1–21.
[4] Michoacan U, L LB, Alvarado-flores JJ, Jes N, Corral-huacuz JC, Aguilera-mandujano A, et al. COVID-19 Pandemic Effect on Energy Consumption in State Universities: Michoacan, Mexico Case Study 2021, **17**, 1–19.
[5] Jiang P, Fan Y Van. Impacts of COVID-19 on energy demand and consumption : Challenges , lessons and emerging opportunities 2021;285.
[6] México. National Registry of Emissions of México-Factor de Emisión del Sistema Eléctrico. 2020.
[7] INECC. Factores de emision para los diferentes tipos de combustibles fosiles que se consumen en México. 2014;52:1–62.

AUTHOR INDEX

Acakpovi, Amevi 100, 112, 142, 163
Adjei, Patrick 112, 142, 163
Afonope, Maxwell 112, 142, 163
Aggrey, George K. 112, 163
Appiah, Frank K 100
Bhusal, Harshvardhan 127
Bondarchuk, R 75
Chanagond, Malati 14, 31
Cuevas-Rodríguez, V D 173
Das, T D 157
Daud, Sarbani 23
Doddi, Suresh 14
Edwards, David J 100
Gadhe, Prakash 127
Gadiya, Prem 127
Girnar, Jwalandhar 127
Govindasamy, Gopalakrishnan 55, 61
Hamidi, Mohd Adnin 23
Hedge, Ramakrishna N 41
Hosseini, Reza M 100
Jadhav, Lalit 127
Jeeragal, Ramesh 14, 31
Kale, Sandip A. ....
Kanasepatil, A B 1
Kinikar, Hemant A 1
Kofi, Daniel 112, 142, 163
Kumar, Atul 48
Kumar, J Praveen 61
Lakyda, I 75
Lakyda, M 69, 75
Lakyda 69
López-Sosa, L B 173
Mamat, Rizalman 23
Manjunath, S 41
Melnyk, O 69, 75
Morales-Máximo, M 173
Offei, Felix 142
Owusu-Manu, De-Graft 100
Pasupula, Ramatejaswi 61
Patil, Amaresha T 31
Puthani, Prashant 14, 31
Rao, K Shubha 81
Rodríguez-Torres, G M 173
Sackey, David M 100, 142
Sahu, Tomesh Kumar 48
Sankaran, Ganesh 89
Sapate, Kumar D ....
Sharma, Rohit 55
Shukla, Pravesh Chandra 48
Slusarchuk, V 69
Soshenskyi, O 69
Sowah, Robert 112, 163
Subramanian, Sunu 55
Sulley, Mustapha 112, 142, 163
Taka, Gem 157
Tay, Godfred 112, 142, 163
Thipse, S S 1
Vasudha, Karanam 81
Vasylyshyn, R 69, 75
Venkatesan, S. 89
Yurchuk, Yu 75